THE ART OF

STARGAZING

I would like to dedicate this book to my daughter Lori,
who continues to be my inspiration in all things
and helps me reach for the stars every day.

THE ART OF
STARGAZING

DR MAGGIE ADERIN-POCOCK

BBC
BOOKS

BBC Books is an imprint of Ebury Publishing
20 Vauxhall Bridge Road,
London SW1V 2SA

BBC Books is part of the Penguin Random House group of companies
whose addresses can be found at global.penguinrandomhouse.com

Penguin
Random House
UK

Illustrations all by Tom Matuszewski
Diagram renditions by Greg Stevenson (pages 2, 7, 10, 18, 24 and 29)

This book is published to accompany the television series *The Sky at Night*,
first broadcast on BBC One in 1957.

First published by BBC Books in 2023

www.penguin.co.uk

A CIP catalogue record for this book is available from the British Library

ISBN 9781785947896

Project Editor: Nell Warner
Contributor: Simon Guerrier
Text design: seagull.net
Jacket and endpaper design: Helen Crawford-White
Production: Antony Heller

Printed and bound in Great Britain by Clays Ltd, St Ives PLC

MIX
Paper from
responsible sources
FSC® C018179

CONTENTS

1. STARS: AN INTRODUCTION 1
THE WONDER OF LOOKING UP 1
WHAT ARE CONSTELLATIONS? 2

2. THE SCIENCE OF STARS 6
WHAT IS A STAR? 6
WHAT ARE STARS MADE OF? 6
STAR FORMATION 8
LIFE CYCLE OF STARS 9
STELLAR RADIATION 13
APPARENT STAR MOVEMENT 14
STELLAR BRIGHTNESS: REAL AND APPARENT 14
EXOPLANETS 16
THE SUN AND ITS SYSTEM 17

3. CONSTELLATIONS: AN INTRODUCTION 21
THE 88 CONSTELLATIONS 21
BAYER DESIGNATION: HOW WE NAME THE STARS 21
CONSTELLATIONS OF THE ZODIAC 23
THE TRUTH ABOUT CONSTELLATIONS 24

4. OBSERVING: AN INTRODUCTION 25
WHERE TO START 25
NAKED-EYE OBSERVATIONS 26
BASIC KIT 27
TROUBLESHOOTING 28
NAVIGATING THE HEAVENS 29

5. THE CONSTELLATIONS 31

FURTHER READING 245
ACKNOWLEDGEMENTS 245
INDEX 246

1.

STARS: AN INTRODUCTION

THE WONDER OF LOOKING UP

Next time you notice it is a clear bright night, wherever you are – in a city or in the countryside – step outside if you can or just get to a window and look up for a few minutes. The view will help put you at peace, and you'll be following in a long tradition that probably spans back to a time before we could communicate with words, but when we could still share the wonder and majesty of the cosmos.

Looking up at the stars has been a constant quest in my life. Someone once commented that even when trying on a VR headset to enter a virtual world, the first thing I do is look up to see the stars.

I spent my teenage years growing up in a council flat in London, but that did not deter me from looking up whenever I got the opportunity. In one of the places we lived, I would run up to the upper floors of the building, where I could look out across the iconic London skyline, serene against a canopy of stars that was only slightly diminished by the city lights.

I remember one particularly clear night, feeling limited by the barrier of the window, I ventured outside for a better view. I stood between the blocks of flats to look up and take in the awe-inspiring vista. When I had had my fill and was starting to feel cold, I turned to go back inside … only to realise that the door had shut behind me, locking me out. It was 3am, I was standing outside in my nightdress and now I was anticipating the wrath of my father when I tried to explain why I had woken him up at such an ungodly hour to let me back in, and why I was half-dressed and standing out in the cold. Luckily, I avoided that conversation as a door to another level of the flats had been left open, so I sneaked back into bed with Dad none the wiser and, once the adrenaline jolt had subsided, I drifted off to sleep with a contented smile on my face.

Since then, I have experienced the joy of working at and visiting some of the largest telescopes in the world. They are generally built on locations atop mountains that give the clearest possible views of the universe.

One of the great joys I currently experience is speaking to school children of all ages. When I first started doing this, I pondered on the hook that would get a 4-year-old excited about space. The answer came to me like a gift:

Twinkle, twinkle, little star,
How I wonder what you are …

We learn this rhyme as children without really thinking about it, but 'wonder' truly is the right word to describe the sky on a clear night, with its thousands of stars glimmering bright in the vast blackness of space. Through the ages, this awesome spectacle has inspired great art as well as scientific inquiry. It has shaped civilisations.

WHAT ARE CONSTELLATIONS?

In ancient times, when people looked up at the night sky, the stars were more vivid than they are today, as there wasn't the light pollution of modern times. People came to recognise patterns in the arrangement of stars. To them, one group of stars might look a bit like a woman seated on a throne, another like a dog or a bear. They wove these characters into the stories they told, interweaving culture with pinpricks of light.

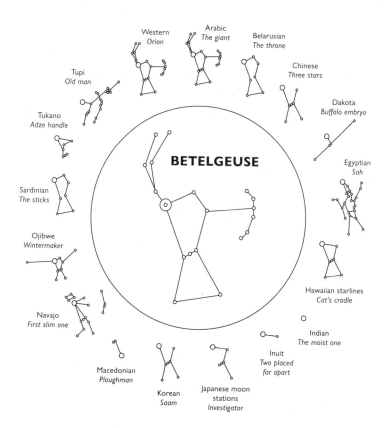

Figure 1: Different interpretations of the stars in **60. Orion**

Today we call these patterns of stars constellations. The word 'constellations' is derived from Latin and means 'set of stars'. This book is a guide to the 88 constellations officially recognised by the International Astronomical Union (IAU). These largely stem from ancient Greek interpretations of the star patterns in the night sky. However, this is just one interpretation of star patterns. Different people and cultures have seen completely different patterns in the stars and related them to their own history, surroundings and stories. Let's take the bright star Betelgeuse; this star is visible at various times of the year in both the northern and southern hemispheres. In the IAU official constellation, it features in **60. Orion**, but that same star, and those around it, have been viewed by other cultures in very varied ways.

We've given each of the 88 official constellations, as recognised by the IAU, its own entry in Chapter 5, and placed them in alphabetical order. Wherever we mention a constellation in this book, it will be numbered and marked – for example, **1. Andromeda** – so that you can look up the full entry and locate it on the star maps at the front and back of the book, and then in the sky.

Let's begin with a very easily recognised pattern of stars. Asterisms, in contrast to constellations, are patterns of stars that seem close together from our perspective on Earth but are often not close together at all. Asterisms tend to be relatively simple shapes consisting of just a few stars, so they're easy to remember. In the northern hemisphere, one of the most widely recognised asterisms is made up of seven bright stars in the constellation **83. Ursa Major**.

In the UK and Ireland, these seven stars have traditionally been called the Plough, as they resemble the ard-type ploughs used to break up soil ready for planting seeds. Such ploughs were once commonly used across the UK and Ireland, so the name of this asterism suggests that people saw something of their own lives depicted in the night sky.

These same seven stars have also been seen as a butcher's cleaver, or as a wagon or 'wain' without its wheels. Since there's a similar, smaller pattern of stars nearby – in the constellation **84. Ursa Minor** – some people (including some in the UK and Ireland) referred to the larger group of seven stars as the Great Wain, while others referred to the larger group as Husband's Wain and the smaller group as Wife's Wain, giving us a sense of the gender politics of the time.

Others linked the seven stars to characters from myth – for example, Arthur's Wain after the legendary king of the Britons, Odin's Wain after the father of the Norse gods, and Göncöl's Wain after a mythological Hungarian shaman. Another name, Charles's Wain, has been linked to the real-life King Charlemagne (c. 742–814 CE), but that name has an earlier derivation from 'churl', an old word for 'husband'. It may be that both meanings apply and the association has changed over time, from generic husband to specific king. Even within the same culture, the meanings of the stars can be fluid.

Different cultures in what are now India, Europe and North America held that these stars form, or are the tail of, a 'great bear' (see **83. Ursa Major** for more on this). In the Inuit tradition, the same stars are seen as a caribou or reindeer. In Burma, stars in this part of the sky are associated with crustaceans (the family of creatures that includes lobsters, prawns and crabs). These are all reflections of the local wildlife that each culture knew.

In Vietnam, the seven stars have traditionally been seen as a rudder; fishing boats still navigate by these stars, so the name seems to come from their usefulness. They're seen as a boat or canoe by other cultures across Asia, perhaps for similar reasons. In China, the same stars make up one of several walls enclosing what's known as the north celestial pole (see page 30), perhaps echoing the great walls built across that country. The stars are also linked to figures from Chinese religion, showing that the same stars can have more than one meaning for the same culture at once.

In the Philippines, these same stars are identified as a traditional kind of pot of cleansing water. In the US, they're a 'big dipper' or ladle. What's more, these seven stars are only visible to observers north of Earth's equator: in the southern hemisphere, they're not seen in the night sky at all, so had no cultural meaning applied to them there in the past.

Of course, there's no correct interpretation of any pattern of stars: one culture's meaning is as valid as any other. But using different patterns and names could make it difficult for astronomers to know which stars or parts of the sky are being referred to. It is helpful to have some consistency, so in 1928 the IAU officially adopted the 88 constellations we detail in this book.

Constellations did more than reflect ancient cultures; they served a practical purpose, too. In recognising these patterns, ancient people could better keep track of the movement of the stars. Many noticed that the stars appear to move across the sky over the course of the night, revolving round a point we now call the north celestial pole. We now know that this apparent movement is because Earth itself is spinning: effectively, we are revolving under the 'fixed' stars.

The patterns that are visible to us also change over a much longer timescale. Some patterns disappear from view for months and then return, in a cycle lasting 365 nights. Ancient people didn't understand the process, that what we can see changes as Earth orbits the Sun over a period we call a year. But they did recognise that it was a regular cycle, and that this knowledge could be useful.

For example, around 4000 BCE, people settled on the banks of the river Nile in what is now Egypt. Every so often, the river flooded and washed settlements away, yet people stayed because when the floodwater receded it left muddy ground that was good for growing crops. Some of these people came to realise that the river flooded around the same time in each cycle of the stars (that is, the floods happened roughly the same time each year). They recognised that the floods coincided with particular stars reaching particular parts of the sky. Soon, they used the stars to identify the best moment to plant or harvest their crops, maximising the amount of food they could grow.

We know something similar happened in many other parts of the world, too, but in Egypt plenty of evidence survives to show us in some detail what happened next – the effect of this knowledge of the stars. Complex mathematics and measuring were needed to work out who owned which bits of land after floods washed away markers and boundaries. Writing helped with labelling plots and boundaries. All these efforts needed to be organised, so there were social structures, hierarchies, government. People could earn food in indirect ways, such as entertaining others with music, dancing, art and stories. Life became richer and something like we know it today because people could recognise the movement of the stars.

Within about 1,500 years of people first settling by the Nile, they had developed the skills and organisation to construct what we think is the world's first large stone building, the Step Pyramid at Saqqara, built as a tomb for King Djoser. As with the pyramids that followed it, this was an awesome demonstration of the power of the kingdom, and it seems that Djoser was in no doubt about where that power came from. Inside the pyramid, his burial chamber was decorated with images of stars, thought to be Polaris (the North Star) and others in the constellation we now call **84. Ursa Minor**. To one side of the pyramid, a mortuary temple was constructed to face due north, again making a link to these northern stars.

Today, thousands of years after Djoser, we're still reaching out to the stars. We use ever more sophisticated telescopes and a range of scientific methods to peer further into the darkness. In the last 30 years, we've discovered thousands of planets in orbit around stars other than our own. We've photographed a star that is 2.5 million years old. We use the movement of the stars to unravel the early history of the universe and predict the future. It's extraordinary, awesome stuff – and all from continuing to wonder at those points of light …

With this book I want us all to share the tranquil joy that is stargazing, to rekindle some of the wonder that our ancestors had for the night sky. I hope it gets you actively looking up at the stars, and helps you be better able to recognise the patterns you see and understand some of the many stories they have inspired. In identifying familiar groups of stars, we can explore the extent of our knowledge to date, while still retaining that sense of wonder.

Now, let's go back to the nursery rhyme 'Twinkle Twinkle Little Star' and see if modern science can answer the question we've wondered about for so long: what the stars actually are.

2.

THE SCIENCE OF STARS

WHAT IS A STAR?

To me, the stars have always been a hope and an aspiration. Since childhood I've dreamed of reaching them, inspired by science-fiction stories such as the *Clangers*, *Star Trek* and *Doctor Who*. But as I seek to get ever closer to them, the question remains: what is a star?

Well, the online encylopaedia Wikipedia defines a star as:

> An astronomical object comprising a luminous spheroid of plasma held together by its own gravity.

But that prompts more questions than it answers, so let's dive in and take a look at the science of stars.

WHAT ARE STARS MADE OF?

The definition above says that stars are made of plasma, but to understand what that is, we need to go back to the ancients again for a moment.

The ancient Greeks believed that things could exist in one of four main states of matter: the 'elements'. The first element was fire, or the imponderable form. The second was air, or the gaseous form. Then there was water, or the liquid form. Finally, there was earth, or the solid form.

The ancient Greek philosopher Anaxagoras (c. 500–c. 428 BCE) believed that the Sun was a ball of metal at a very high temperature – whether it was solid or liquid would depend on what metal it was made of. Now, this was a bit wide of the mark, but he also believed that the Moon was made of similar matter to Earth and that it only shone because it reflected light from the Sun, which are amazing deductions considering what was known at the time, and which have since been shown to be true.

Aristotle (384–322 BCE) added a fifth element to the first four, 'ether', later known as 'quintessence' ('quint' meaning 'five' and 'essence' meaning 'element'). It was thought that this fifth element was the refined, pure essence of all the other elements and that it flew upwards and outwards at the time of creation, to fill the space beyond Earth and hold the stars and other planets in place – but that didn't tell us what kind of metal (or anything else) the Sun was made of.

By the nineteenth century, we had gained a better understanding of thermodynamics (the behaviour of heat), and the idea that the Sun was a huge burning lump of coal or a ball of hot gas did not stand up to scrutiny.

People knew that the Sun produces vast amounts of energy, but they struggled to explain how.

The next big step forward was triggered by a physics conference on 'Stellar Energy Generation' in 1938. Attendees were given the density, temperature and chemical composition of the Sun and set the challenge of devising a process that would generate energy. German-American physicist Hans Bethe (1906–2005) had not originally wanted to attend the conference, but he soon took on the challenge. Bethe eventually worked out the fusion pathways that power the Sun, which involve the synthesis of elements from helium to iron. For this extraordinary work, Bethe was awarded a Nobel Prize in Physics in 1967 for 'his contributions to the theory of nuclear reactions, especially his discoveries concerning the energy production in stars' (we'll explore this more in 'Life Cycle of Stars', below).

Today, we also define four states of matter. We are familiar with the first three – solid, liquid and gas – which overlap with the ancient definitions. We have ditched the fifth element (quintessence/ether), but strangely enough, fire has sort of remained.

A burning candle was the kind of fire people in ancient times understood. The temperature in the hottest part of a candle flame is around 1,100°C and that flame is a gas. However, we now know that if a flame has a temperature above 3,000°C, it could be in the fourth state of matter, known as plasma, which has some very strange properties.

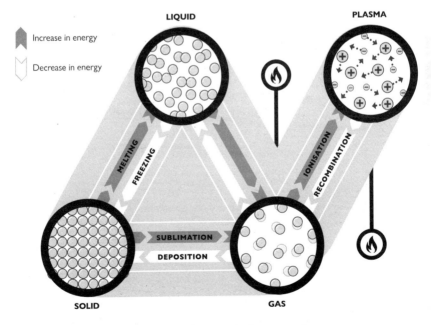

Figure 2: Interactions between the four states of matter

In all other states of matter – solid, liquid and gas – the negatively charged electrons associated with each atom remain tethered to their positively charged nucleus. Yet in a naturally occurring plasma, the temperatures are high enough that this no longer holds true. All or some of the electrons associated with an atom can be stripped from the atom and then wander around freely in a sort of gaseous soup embedded with positively charged nuclei, called ions.

Artificial plasmas are generally generated by electrical and/or magnetic fields being passed through a gas. But the plasma is only maintained while these fields are applied. We encounter these plasmas in our everyday lives in the form of fluorescent lamps, LED screens and neon signs. When an electric current is passed through the low-density gas contained in a fluorescent tube, temperatures are reached that ionise the gas, liberating the electrons and creating a plasma – the light in the display.

Naturally occurring plasmas on Earth include lightning and the aurora borealis (or Northern Lights). Stars are huge, bright balls of plasma, where temperatures run into the millions of degrees Celsius, rather than the mere thousands.

STAR FORMATION

Stars are formed inside vast clouds of gas and dust called nebulae (a word meaning 'clouds'). Nebulae are enormous: one of the largest discovered to date is the Tarantula Nebula in **33. Dorado**, which stretches for 1,800 light years. In other words, it is 17,029,314,850,645,440km across and it would take a beam of light 1,800 years to travel from one side to the other!

Many nebulae are formed from or contain the remnants of stars that existed previously but exploded in a process called a supernova. This releases elements that help create new stars. The Crab Nebula in **78. Taurus** sits 6,535 light years away from Earth. It formed as the result of a star going supernova, which was seen from Earth in 1054. Supernovae are so incredibly bright that even though this event happened 60 thousand trillion kilometres away – that is, 60,000,000,000,000,000km! – it was witnessed on Earth and recorded by Chinese astronomers.

Messier Objects
The Crab Nebula is the first of 110 nebulae and star clusters in the catalogue begun by Charles Messier (1730–1817), and so it is also known as M1. Messier was a French comet hunter in the 1700s. He compiled a list of hazy objects that could be mistaken for comets. These fuzzy objects are now known to be clusters, nebulae and galaxies. Because he was observing from France, he couldn't see far enough south to the constellations that circle the south celestial pole.

Nebulae are usually cold and stable, but if disturbed externally (such as by a nearby supernova) or internally (such as by turbulence from deep within the nebula), they can form high-density regions called knots. These knots contain sufficient gravity that, over time, the gas and dust gather together around them, making them an ever denser and hotter cloud. As this process continues, the cloud starts to spin around a dense centre. As the cloud spins, it flattens out into what is called an accretion disc around the dense centre. Then, as more material is attracted to the centre, the pressure and temperatures in the dense core increase and a protostar is formed. Material left in the disc will condense to form planets around the new star. When the star's innermost temperature reaches around 10 million kelvin,* a process begins called nuclear fusion – which we'll come back to shortly.

Often, several new stars are born in the same nebula. Open clusters can contain anywhere from 10 to 10,000 loosely packed young stars that are often bright and blue. Examples visible to the naked eye include the Pleiades and Hyades clusters near **78. Taurus**, Praesepe (also known as the Beehive Cluster) in **12. Cancer** and the Double Cluster in **63. Perseus**. We think our Sun was once part of such a cluster, but clusters often disperse over billions of years.

> **Orion Nebula**
> The Orion Nebula in **60. Orion** is a stellar nursery where many new stars are forming. It is also one of the few nebulae visible to the naked eye. See if you can check it out on a clear night.

LIFE CYCLE OF STARS
We have addressed how stars are formed, but how do they live and die? Fusion is a process whereby nuclei within the plasma of the star fuse together to make new elements – for instance, turning two hydrogen nuclei (H) into a helium nucleus (He):

$$H + H \rightarrow He + ENERGY$$

A small amount of mass is lost in this process, but this small amount of mass is converted into a huge amount of energy, as shown in the simplified equation above. The process of mass conversion is governed by Einstein's famous equation:

$$E = mc^2$$

where E is the energy released, m is the mass lost in the conversion and c is a constant, a scaling factor, which is the speed of light. The speed of light c is equal to around 300 million metres per second, and in this equation it is

---

* The kelvin is the primary unit of temperature used by scientists. 1K = −273.15°C.

squared, so a small amount of mass releases huge amounts of energy. This is the reason stars shine so brightly: the energy from the fusion process is released as different types of electromagnetic radiation, which we can feel and detect here on Earth.

During the lifetime of a star, a balance of forces keeps the star stable. This balance is between the fusion process, which causes an outward force from the centre of the star, and the inward pull of gravity towards the centre of the star. This balance continues throughout the lifetime of the star, until it runs out of fuel.

Hot, blue stars use up fuel relatively quickly, but older, red stars burn fuel more slowly. Most stars follow the same trend over time. By comparing different stars and grouping them into types based on colour and temperature, we've been able to deduce the ordinary life cycle or 'main sequence' of a star. That also means that we know which stars are especially unusual.

Our Sun is on the main sequence. It is now about 4.6 billion years old, which means it is in the middle of its life and has enough hydrogen and helium to fuel it for another 4.5–5.5 billion years.

Astronomers have seven main classes of stars:

- Blue/white O-type stars have high surface temperatures of about 30,000K. They emit most of their light in the ultraviolet part of the spectrum. When that light is diffracted (see 'Stellar Brightness: Real and Apparent', below), we see strong absorption lines (see page 15)

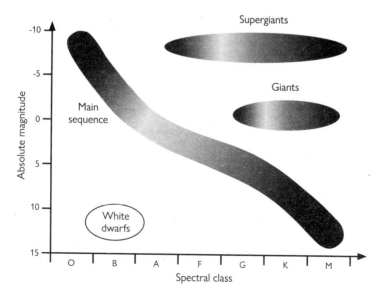

Figure 3: A simplified Hertzsprung–Russell diagram

corresponding to helium. An example is the star ζ Orionis or Alnitak, in the constellation **60. Orion**.

- Bright blue B-type stars have surface temperatures between 10,000 and 30,000K and range from being 2 to 16 times as massive as our Sun. Such stars include α Leonis or Regulus in **46. Leo**.
- Cooler, white A-type stars have strong hydrogen absorption lines and include α Lyrae or Vega in the constellation **52. Lyra**.
- Yellow-white F-type stars range in temperature from 6,000 to 7,500K.
- Yellow G-type stars such as our Sun have surface temperatures of about 5,500K and many absorption lines.
- Pale orange K-type stars range in temperature from 3,700–5,500K, have very weak hydrogen absorption lines and are on average about half the mass, radius and brightness of our Sun.
- Red M-type stars, including α Orionis or Betelgeuse in **60. Orion**, are cool enough at about 3,000K for molecules such as titanium oxide to form in their atmospheres, giving complex spectra.
- Beyond M, there are some cooler stars classed L and T.

For most of a star's life, the fuel it uses is hydrogen to make helium. But as the hydrogen runs out, the star's gravitational force starts to dominate, putting pressure on the core and increasing the temperature. The increased temperature means that the fusion of helium can now start, producing heavier elements. For example, two helium nuclei will form beryllium. Beryllium and helium can then form carbon. Carbon and helium can form oxygen. (note: γ = light)

$$^4He + {}^4He \leftrightarrow {}^8Be$$

$$^8Be + {}^4He \rightarrow {}^{12}C + \gamma$$

$$^{12}C + {}^4He \rightarrow {}^{16}O + \gamma$$

$$^{16}O + {}^4He \rightarrow {}^{20}Ne + \gamma$$

What happens when stars run out of fuel depends on their size. The higher the mass of the star, the heavier the elements it can create in its core. This is because heavy-element fusion requires higher temperatures, which only the most massive stars can attain.

For stars the mass of our Sun, running out of fuel is where the fusion process ends, as the gravitational collapse of the star is not enough to increase the core temperature sufficiently for further fusion into heavier elements to take place.

Stars up to about eight times the mass of our Sun gradually get bigger, redder and brighter, until they become red giants, with outer layers some ten times bigger than their original size and moving well off the main sequence. Examples include α Boötis or Arcturus in the constellation

9. Boötes. Although the outer shell of these stars has expanded, their inner cores continue to collapse and thus increase in temperature, which enables further fusion to take place.

After a billion years, when it has exhausted this fuel, the star sheds its outer layers until all that is left is a very hot core surrounded by what is called a planetary nebula – a cloud of gas that gets its name because it looks, when viewed through a small telescope, a bit like a planetary disc and was once thought, wrongly, to be the birthplace of planets such as Uranus and Neptune. Over time, this hot core will cool and fade, becoming a small, dense white dwarf star. White dwarfs are too faint to see with the naked eye, but viewed through a telescope the bright star Sirius in the constellation **14. Canis Major** proves to be two stars: one an A-type main sequence star, and the other the brightest known white dwarf.

For stars with masses around nine times that of our Sun, the depletion of helium as a fuel causes further gravitational collapse and an increase in core temperature/pressure that enables carbon to fuse to form elements such as magnesium, sodium and aluminium. This process of making heavier elements continues until iron is made through fusion at the heart of the star. These stars become red supergiants, such as α Orionis or Betelgeuse in **60. Orion**.

The process of the formation of elements heavier than iron remained a mystery until the work of astrophysicist Margaret Burbidge. In 1957, she was the lead author of a game-changing scientific paper credited to 'B²FH' (the initials of authors Margaret and Geoffrey Burbidge, William Fowler and Fred Hoyle) and titled 'Synthesis of the Elements in Stars'. This paper states that when stars die their contents are released into the stellar medium from which new stars are made.

Margaret Burbidge

Although Margaret Burbidge made a major contribution to our understanding of stars, she faced difficulties in her early work. As an experimental astrophysicist, she wanted time on telescopes to make observations, but as a woman working in the 1950s, she was denied access. To get round this, her husband, a theoretical astrophysicist, would apply for time on the telescopes and then pass this on to Margaret so she could do her work!

When a massive star – more than ten times the mass of our Sun – runs out of fuel, the gravitational collapse of the star increases the temperature and pressure in the star's core to such high levels that in those last few moments fusion can produce the heaviest elements, including silver, gold and platinum. The fusion process that causes the production of these heavy elements counteracts and overcomes the gravitational collapse of the star.

The star becomes unstable and the outward force explodes the star in an event called a supernova, releasing all the star's fusion products out into the stellar medium, to be incorporated into another star in the future.

Such a supernova can leave behind a tiny, dense and fast-spinning neutron star, or in some cases a black hole – a core so compact and dense that even light cannot escape its gravity. Neutron stars and black holes cannot be seen with the naked eye; indeed, black holes can't be seen directly. Instead, we deduce their existence from the observed effects of their enormous gravitational force.

The death of older stars provides the fuel and heavy elements for the next generation of stars in an ongoing life cycle, and we are the direct beneficiaries of this extraordinary process. All the elements in our bodies were forged in the heart of a star. The heavier elements may have passed through a number of stars before becoming part of us. So not only are we stardust, but that dust may have passed through a number of different stars before settling in us!

STELLAR RADIATION

As discussed in the last section, the brightness of a star is the result of the nuclear fusion reactions that are taking place in its core, converting hydrogen to helium (and, in larger stars, to other elements, too). In the process, huge amounts of energy are produced that radiate out in all directions.

Our nearest star, the Sun, produces energy that makes life on Earth possible: it keeps us warm, and plants convert sunlight into sugars through a process called photosynthesis. This process sits at the base of the food chain that we and other animals depend on.

The energy from the Sun comes to us as different forms of radiation (see Figure 6, page 29). Visible light is the electromagnetic radiation that we detect with our eyes that allows us to see things. Infrared radiation warms us and the planet. Ultraviolet radiation can burn and damage our skin and the cells in our bodies. X-rays and gamma rays are very high-energy electromagnetic radiation that can cause grave damage to living cells – although this radiation is produced by the Sun, our atmosphere is opaque to it, so we are protected from its harmful effects.

We have developed technology to directly harness the Sun's radiation in the form of solar power. But the fossil fuels that we use are also derived from the Sun's radiation. Coal, oil and gas are the remnants of plants that lived millions of years ago and that trapped solar energy, which we release by burning them. Almost all the energy we use here on Earth has its origins in solar energy. The only two that I can think of that don't stem from solar are nuclear and geothermal energy.

This radiant energy from the Sun – and from all stars – travels at the speed of light: 299,792,458m/s or 1,080,000,000km/h. But space is so vast that, even at that speed, it takes time for the light from any star to reach us here on Earth. For example, the Sun is an average of 150 million kilometres from Earth and it takes eight minutes for all the different types of radiation including visible light to travel that distance.

The next nearest star to us is Proxima Centauri, which is one component of the triple star system Alpha Centauri; these three stars are found in the constellation **19. Centaurus**. It takes light from these stars 4.28 years to reach us because they sit 4.28 light years away. Light from other stars that we can see with the naked eye can take hundreds or even thousands of years to reach us. As a result, some of the stars we observe today may no longer exist, as they will have died in the time it took their light to reach us. In this way, stargazing is a form of time travel. When we observe objects out in space, we see them not as they are right now, but as they were when the light we see left them: we are looking back in time.

Using telescopes, we can see more distant objects – and so look further back in time. In 2018, NASA's Hubble Space Telescope photographed an enormous blue star known as Icarus in the constellation **46. Leo**, the most distant individual star ever seen. The light we see left Icarus more than 9 billion years ago, before Earth even existed.

APPARENT STAR MOVEMENT

Stars appear to move through the night sky. Over the course of a night, they seem to move in slow arcs, circling in an anticlockwise direction as seen from the northern hemisphere (they move clockwise in the sky in the southern hemisphere). In the northern hemisphere they appear to rotate round Polaris, the North Star or Pole Star that sits in **84. Ursa Minor**. In the southern hemisphere, there is no pole star at the centre of rotation, but the stars still seem to rotate around a common point. But it is not the stars that are moving across the night sky; we are the ones moving. As Earth rotates on its axis once every 24 hours, the stars appear to sweep around our axis of rotation, but it is us rotating, not the stars.

As Earth orbits the Sun over the course of the year, we also lose sight of 'seasonal' stars for a while.

But we're not the only ones moving: the stars themselves are drifting through space, some moving together, others further apart. Given the vastness of space and the distance we're watching from, that movement is barely perceptible, but we have to update our star maps with small changes every 50 years (the next update is not due until 2050).

STELLAR BRIGHTNESS: REAL AND APPARENT

Some stars appear bigger and/or brighter than others, which we indicate in the illustrations in this book by using spots of different sizes. Astronomers speak of stars' 'apparent brightness' and rank them in order of 'apparent magnitude': the lower the figure, the brighter the star. For example:

−26.7: The Sun (so bright it prevents us from seeing most other stars!)

–1.4: Sirius, in the constellation **14. Canis Major**; the brightest star in the night sky

0: Vega, the brightest star in the constellation **52. Lyra**

3: About the limit at which we can see stars with the naked eye from light-polluted places such as London

6: About the limit at which we can see stars with the naked eye anywhere

31: The faintest stars detected by the Hubble Space Telescope

But this is only the stars' apparent brightness – how bright they appear as seen from Earth. In fact, a dimmer-seeming star may actually be brighter but much further away than one that appears brighter to us. So astronomers also speak of 'absolute magnitude', or what the apparent magnitude of stars would be if we saw them all at the same distance of ten parsecs (309 trillion kilometres).

We can deduce the absolute magnitude of some stars quite easily. For example, Cepheid variables (see **20. Cepheus**) are large, bright stars whose brightness varies over regular periods of 1–70 days. We know how that period is related to their absolute magnitude. What's more, by comparing the absolute magnitude to their apparent magnitude we can work out how distant they are from Earth.

Another useful tool for understanding the true brightness of stars is spectroscopy. This is a technique that I used throughout my PhD and also while working on ground- and space-based telescopes.

Spectroscopy is a technique whereby you take the light from an object and then use various optics to spread that light into its component colours. The technique is similar to the process that forms rainbows. Raindrops contain water that disperse sunlight into its component colours, as the different colours (or wavelengths) of light travel at different speeds through water. If this technique is used on astronomical objects, analysis of the spectra (the resulting rainbow array of colours) can tell us a lot about the object under observation. For instance, small black bands that appear across the spectra are known as absorption lines (or absorption bands) and can give an indication of what chemical reactions/fusion processes are happening within the star. Absorption lines occur because of the absorption of energy due to different energy states in an atom or molecule. The absorption lines for different atoms and molecules have a distinctive 'fingerprint', so by looking at the absorption lines signature, it's possible to analyse the chemical composition of the object. If we can get a spectral signature of a star, we can deduce the fusion processes happening within it, and then we can have a good guess at the size and probable brightness of the object. This, compared with its apparent brightness, can give us an indication of how far away the star is from us.

But absorption lines can tell us even more about the object under observation. Shifts in the absorption signature can tell us about the movement of the object either towards or away from us. If the fingerprint is shifted towards the blue end of the spectrum, then the object is moving towards us. If it is shifted towards the red end of the spectrum, then the object is moving away from us. It was spectral analysis of this kind that enabled us to work out that the universe is expanding, as nearly all objects have spectra that indicate they are moving away from us.

With imagination, we can track this process backwards to when, billions of years ago, the whole universe was a single point that expanded outwards. This idea is known as the Big Bang – and this sense of the earliest moments of our universe comes from looking at the stars as they are now.

EXOPLANETS

Looking at the light emitted from stars can also give us an insight into one of the most exciting phenomena in modern astronomy. Regular dips in the light emitted by a star can tell us a few things about the star, but sometimes the dips indicate that there is an exoplanet – i.e. a planet orbiting a star other than our Sun; 'exo' meaning 'outside' the orbit of the Sun – orbiting this distant star. This has been a major leap forward in astronomy.

For centuries people suggested that there might be exoplanets out there, but they are very hard to spot. The difficulty lies in the fact that, as we have discussed, stars produce vast amounts of radiation that we can detect with our eyes, binoculars and telescopes, while planets produce virtually no radiation at all. In our Solar System, we can see the eight planets orbiting the Sun because they reflect the Sun's light back towards us. This means that if you go outside tonight and see any of the planets that can be spotted with the naked eye, what you are actually doing is detecting light with your eye that was generated in the heart of the Sun via a fusion process. That light then left the Sun and travelled many millions of kilometres through the vacuum of space to hit the planet you are observing, reflected off it, then again travelled through the vacuum of space to end up in your eye. That is quite an epic journey!

However, the planets reflect only a tiny amount of light compared to the brightness of the Sun. So even if we are looking at a planet in our next-door star system, such as Proxima Centauri b in **19. Centaurus**, a mere 39.7 trillion kilometres away from us, the amount of light reflected by the exoplanet is just too dim for us to detect.

This is where the detective work comes in. If we can't pick up the light reflected by the planet, what else can we do? Well, there are two main methods for detecting exoplanets. The first uses the fact that as an exoplanet orbits its star, the star and planet have a gravitational pull on one another that can be detected as a wobble in the movement of the star. This technique is called the radial velocity method. By careful analysis of the wobble, we can work out the mass of the exoplanet. But this technique is only effective for the detection of large planets that don't sit too far from their star. One of the goals of exoplanet detection is to find planets similar to Earth, which is currently beyond the capabilities of this technique.

The second technique is called the transit method. This detects a dip in the light coming from a star, because when an exoplanet travels between us and the star, a tiny amount of the starlight is blocked. For this method to work, the planet, the star and our observation point all have to be at the right orientation so that the planet passes between us and the star.

As you can imagine, the stars that we see in the night sky are tiny points of light, so monitoring this tiny dip in the light output is very challenging. But it's well worth persevering because with the transit method you can find out a lot about an exoplanet.

The period of the exoplanet's orbit (i.e. the time the planet takes to go around its star) can be calculated based on the frequency of the repeated dips in starlight. The size of the planet can be inferred from the drop in the brightness of the star. We can even deduce when there are multiple planets in orbit round the same star. The planet 55 Cancri e is one of at least five planets orbiting a single star in the constellation **12. Cancer**.

Another more recently developed calculation using the transit method really blows my mind. If we, the planet and the star are all at the correct orientation, then the chemical composition of the planet's atmosphere can be understood. This is done by measuring incredibly small changes in the star's spectrum when some of the starlight passes through the very thin band of atmosphere that surrounds the planet. This technique requires a lot of skill, but we have found Earth-sized exoplanets with water vapour in their atmospheres. This opens up the very exciting prospect of one day possibly finding signs of life out there.

To date, around 5,000 exoplanets have been spotted, but as technology evolves I am sure that number will grow. I am very proud to have been one of the scientists that has worked on the James Webb Space Telescope. This is the largest space telescope ever built and one of its primary goals is to look for exoplanets and analyse their atmospheres. Launched on Christmas Day 2021, the James Webb Space Telescope is poised and ready to change our understanding of the universe.

We are still looking and learning. But of course the star system we know most about is the one we're part of …

THE SUN AND ITS SYSTEM

In comparison to other stars, our Sun is fairly average and almost boring. Yet it sits at the heart of an extraordinary, diverse collection of objects, all caught in the Sun's powerful gravitational force. When I speak to kids about this, I call it the Gravity Gang, but its official name is the Solar System.

The Sun itself – also known as Sol – is by far the largest and most massive object in the Solar System. It contains about 99.8% of the mass of the whole system. The 0.2% that is left over is divided between the eight planets and their satellites, the five dwarf planets, the comets and asteroids, and the dust and gases that surround the Sun.

We've looked at the processes that happen in the heart of stars. Now let's look into the heart of an average star to see what is going on.

Just like Earth, the Sun is made up of multiple layers. Surprisingly, the outermost layer, called the corona, has a higher temperature than the layer below, the chromosphere. Temperatures in the corona sit at 1 million degrees Celsius. A NASA spacecraft called the Parker Solar Probe has passed through the outer reaches of the Sun's corona and lived to tell the tale. One of the questions it hopes to answer is why the corona is so hot.

As we are discovering other exo-solar systems, it is a good time to contemplate our own. This book's main focus is on the constellations and stargazing, but many of the planets in our Solar System are visible unaided, so here is a little more about them, including how best to see them.

The four innermost planets – Mercury, Venus, Earth and Mars – are all rocky worlds, but that is where the main similarities end.

Mercury is small and heavily cratered (it looks very much like the Moon), and it always presents the same side to the Sun. As a result, that side is very hot: about 450°C at the equator. By contrast, the far side is in permanent shadow and is a chilly −180°C. A European Space Agency spacecraft,

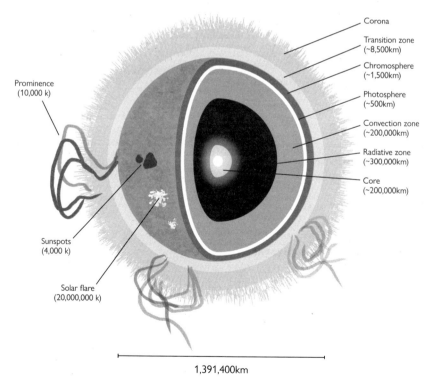

Corona

Transition zone
(~8,500km)

Chromosphere
(~1,500km)

Photosphere
(~500km)

Convection zone
(~200,000km)

Radiative zone
(~300,000km)

Core
(~200,000km)

Prominence
(10,000 k)

Sunspots
(4,000 k)

Solar flare
(20,000,000 k)

1,391,400km

Figure 4: A cross-section of the Sun

BepiColombo is due to enter orbit round Mercury in 2025. Mercury can be seen with the naked eye, but it is hard to spot as it sits so close to the Sun. It can usually be spotted about 90 minutes before sunrise or 90 minutes after sunset. Check astronomical calendars to see when it is visible.

Venus is more distant from the Sun than Mercury, which should make it cooler, but thick clouds of concentrated sulphuric acid and an atmosphere of carbon dioxide stop heat from escaping and raise the average temperature to 462°C. That's hot enough to melt lead! We often see Venus in the sky without knowing it. It is the third brightest object in the sky after the Sun and the Moon. As with Mercury, Venus can be seen just before sunrise or just after sunset, which is why it is sometimes known, misleadingly, as the morning or evening star (although it's not a star; it's definitely a planet!). Both Mercury and Venus have phases like the Moon, but these can be hard to spot without a telescope. Venus is a wonderful planet to see; it looks truly magnificent in the dimming or brightening sky.

Earth is roughly the same size as Venus, but it is very different: its atmosphere warms the planet just enough for there to be liquid water present on the surface. That water is essential to the abundant, diverse life here – the only place in the whole universe where we know life exists. The surface of Earth is also very varied, including oceans, land masses, ice caps and arid deserts. Tectonic plates mean Earth's surface is constantly in flux, leading to earthquakes and volcanoes. The fact that we live here means we can sometimes take for granted the diversity of life and what a spectacular, special planet this is!

Mars is by far my favourite planet out there (second only to Earth), as I think that it has the potential to be a home to humankind in the future, though it is a small, dry place just over half the size of Earth. Not too surprisingly, it's colder than the other inner worlds as it is that much further away from the Sun: it has an average temperature of −58°C at the equator, rising to about 0°C at noon (not that different to temperatures experienced in Antarctica). As on Earth, there are ice caps and some evidence of water. We've sent many probes to Mars hoping to find evidence of life – current or ancient – but nothing has been found … yet! Mars is relatively easy to spot in the night sky if one is given guidance (via an app or online instructions), and it has the added bonus of having an orange/red hue, even when seen with the naked eye.

Beyond Mars are two enormous worlds of gas. Jupiter is so big that a storm in its upper atmosphere is the same size as Earth! The storm, called the Great Red Spot, has been raging since at least 1830. Saturn is a little smaller than Jupiter, but still huge. Weather on both these planets includes rain made of liquid diamonds due to the very high temperatures and pressures experienced in their atmospheres. Both these planets are clearly visible with the naked eye, and a telescope will reveal their larger moons, faintly. If you make this observation, you are following in the footsteps of astronomical legends such as Galileo Galilei.

Beyond Jupiter and Saturn are two slightly smaller worlds of gas, the so-called ice giants, Uranus and Neptune. Neither of the ice giants are visible unaided, and a fairly large telescope is needed to do them justice. Uranus is unusual in that, compared to the other planets, it rotates on its side. We also know that it smells of rotten eggs because its cloud tops are full of the same chemical that makes decomposing eggs smell: hydrogen sulphide.

The average surface temperature of Neptune is −214°C, which makes it the coldest planet in the Solar System, and yet, given its distance from the Sun, it should be colder still. We think it must have some internal heat source, but we don't yet understand the process. It also has winds of 2,000km/h.

The four outer planets all have ring systems of ice and rock in orbit around them. Saturn's are the brightest and best known. All the planets except Mercury and Venus have moons: Earth has one, while Saturn has at least 82. These moons can be as diverse as the planets. Jupiter's moon Europa has a crust of ice covering a liquid saltwater sea, and Io is the most volcanically active body in the Solar System. By contrast, Saturn's moons Enceladus and Titan, and Neptune's moon Triton have volcanoes of ice.

As well as the eight planets, there are five smaller dwarf planets. The most famous of these is Pluto; its surface is largely frozen nitrogen with huge mountains of water ice. In addition, the Solar System is home to thousands of asteroids, comets and meteors, as well as gas and dust, all in motion, orbiting the Sun in paths that range from neat almost-circles to much more eccentric ellipses.

Beyond the planets, there are ice-rock objects arranged in a vast torus or doughnut shape that we call the Kuiper Belt. Beyond that, we think the Oort Cloud contains billions of ice-rocks, some of which sometimes fall inward towards the Sun and become comets.

We still have lots to learn about the outer fringes of our Solar System. Our furthest-reaching space probe is Voyager 1, launched in 1977. After 45 years it is now about 23 billion kilometres away, or about 155 times the distance from the Sun to Earth. Yet it still has a long way to go before it escapes the Sun's gravitational influence, which extends a phenomenal 100,000 times the distance from the Sun to Earth. Who knows what else we'll find in it?

The Solar System, then, is a huge, rich, varied, busy community. It may be that other star systems are just as rich and varied as ours − we're just beginning to find out …

3.

CONSTELLATIONS: AN INTRODUCTION

THE 88 CONSTELLATIONS

In 1928, the International Astronomical Union (IAU) defined 88 patterns of stars, the officially recognised constellations that are the subject of this book. There is an entry for each of them in Chapter 5.

These 88 official constellations are largely based on long-recognised patterns, such as the 48 constellations of the ancient Greeks, which had probably been around for centuries before they were written down in *Phenomena*, a poem by Aratus (c. 315–240 BCE). The official constellations also incorporate the 12 constellations of the astrological zodiac (see page 23), as well as more modern patterns.

In one case, an ancient pattern, "Argo Navis", was broken up into three to fit the new system – see **17. Carina**, **68. Puppis** and **85. Vela**. Other familiar patterns were defined as part of a larger constellation – for example, the seven stars of the Plough (see Chapter 1) are defined as part of **83. Ursa Major**.

In fact, the IAU defined not only the patterns of stars but also the boundaries between them, dividing the entire night sky into 88 interlocking pieces. As a result, every star in the sky is in a constellation, even if they're not one of the dots that make up the outline of the character, animal or object that gives the constellation its name. This means that any new discovery of a star (or other celestial object) easily fits into the system.

Under the IAU system, we also know what to call these new stars.

BAYER DESIGNATION: HOW WE NAME THE STARS

The year in which the 88 constellations were officially recognised, 1930, was a big year for astronomy. For some time before this, telescopes and photography had been used to study the night sky in more detail, identifying ever more stars and other objects. Then, on 18 February 1930, 23-year-old Clyde Tombaugh spotted a discrepancy between two photographs of the same part of the sky taken on different nights the month before. He soon confirmed that one speck of light had moved position between the two images, meaning that it was not a faint star but a previously undiscovered planet, in a very distant orbit round our Sun.

The discovery made news across the world, and one 11-year-old girl in Oxford, England, was struck by the fact that this 'new' planet hadn't yet been given a name. Venetia Burney knew that the other planets in the Solar System were named after gods from classical mythology, so she suggested that the faint, distant world should be named in the same way, and she nominated Pluto, the god of the underworld. Other names were suggested and argued over, but Venetia's suggestion caught on and the name Pluto was formally adopted on 1 May 1930. (As a result of further discoveries of similar distant objects, in 2006 the IAU redefined Pluto: it's no longer a planet but one of five dwarf planets.)

But newly discovered stars couldn't be named in this way. To understand why not, let's again use the example of the Plough, the seven bright stars in the constellation **83. Ursa Major** that have been recognised by different cultures since ancient times. If, using a telescope, we spotted a previously undiscovered star near the Plough, how would we decide what to call it?

As with Pluto, we might be guided by the names already given to known bodies in the same group. Whereas (in the Western tradition) the planets of the Solar System are named after gods of classical mythology, we tend to know stars by the Arabic names given to them by medieval Islamic astronomers. These astronomers knew the brightest star in the Plough as *żahr ad-dubb al-akbar*, or 'the back of the bigger bear', which is usually shortened to Dubhe (the 'bear'). The second brightest star in the Plough is Merak, meaning the 'loins' of the bear – the part of its body just below the ribs. So to name our new star, we might look for another Arabic word to describe a different part of the bear's body.

However, the new star might not correspond to a particular body part, or it might be located outside but close to the stars that make up the outline of the bear. We would need complicated names to describe this, such as 'upper part of the bear's left front thigh' or 'a bee buzzing near the bear's right shoulder'. If we followed this convention, the more stars we discovered, the more convoluted the descriptions we would need.

As we've seen, at the time the IAU agreed upon the 88 constellations, telescopes and photography were being used to identify ever more stars. A notable example took place in 1925, when astronomer Edwin Hubble settled a long-standing argument about the object M31, which to the naked eye is a cloudy smear in the constellation **1. Andromeda**. Some believed this to be a nebula of gas where bright, new stars could be being born. Hubble used Cepheid variable stars (see **20. Cepheus**) to show that M31 was more distant than previously thought – in fact, it was staggeringly so. We now think M31 is some 23,651,826,000,000,000,000km away and that its light takes 2.5 million years to reach us.

That we can see this extraordinarily distant object unaided means it must be enormous. We now know that M31 is about 2,100,000,000,000,000,000km in diameter, that light takes some 220,000 years to travel from one edge of it to the other, and that it is composed of some 1,000,000,000,000 (1 trillion) stars arranged in a barred spiral. It is a galaxy, the Andromeda Galaxy – the first one to be recognised other than our own.

We've found many more galaxies since 1925. In 2021, from data collected by the New Horizons space probe, it was estimated that there are 200,000,000,000 galaxies within range of our telescopes, and each galaxy contains trillions of stars. That's a lot of stars, and it simply wouldn't be practical to argue over separate names for them all.

That's why the IAU's system of 88 constellations is a great help. The name of each constellation has two forms: the nominative, which is the name of the constellation itself in Latin (e.g. Ursa Major, 'Great Bear'), and the genitive or possessive name (e.g. Ursae Majoris, 'of the Great Bear').

In principle, all the stars in a constellation are named in order of brightness using the Greek alphabet. The brightest star in **83. Ursa Major**, previously Dubhe, is now officially Alpha (or α) Ursae Majoris ('the alpha [or first] of the Great Bear'). Merak, the second brightest star, is now Beta (or β) Ursae Majoris, and so on. The names can also be abbreviated: α UMa, β UMa, etc. Any newly discovered star can be added to the system, and the name it is given is known as its Bayer designation, because the system was first used by German astronomer Johann Bayer in his star atlas of 1603. It has been in use for such a long time that there are a few discrepancies – for example, where a beta star is actually brighter than the alpha – which may be because better observations have been made since the designations were applied, or because the boundaries of the constellations have changed and what was once a bright star in one constellation is now a not-so-bright star in another. We'll address these oddities in the relevant constellations in Chapter 5.

The old names, such as Dubhe and Merak, are still often used, so where relevant we'll use both here, too.

CONSTELLATIONS OF THE ZODIAC

Outside astronomy, the word 'constellation' is often used to refer to the 12 patterns of stars in the astrological zodiac. With a little adaptation, these have been incorporated into the system of 88 constellations recognised by the IAU. So what is the zodiac?

The eight planets of the Solar System all orbit the Sun in roughly the same plane. A result of this is that, seen from Earth, the other seven planets always appear within a band 8° above or below the ecliptic, an imaginary line that traces the apparent path of the Sun across the sky. The zodiac is this 16° band around the ecliptic in which the planets appear.

Over the course of a year, the Sun apparently passes in front of the same patterns of stars. The astrological zodiac divides these stars into 12 distinct patterns or signs. In astrology, the sign of the zodiac at the time of your birth is said to correspond to your personality, while the movements of objects through the night sky are thought to mirror – and even foretell – events here on Earth.

We make no such claims of the zodiac in the IAU system, which added a thirteenth constellation, **59. Ophiuchus**, so that the Sun takes exactly four weeks to pass in front of each one. In the IAU system, the zodiac consists of:

  4. Aquarius
66. Pisces
  7. Aries
78. Taurus
38. Gemini
12. Cancer
46. Leo
86. Virgo
49. Libra
73. Scorpius
59. Ophiuchus
72. Sagittarius
16. Capricornus

THE TRUTH ABOUT CONSTELLATIONS

Although we see a constellation as a group of stars, they only appear to be close to one other as seen from Earth. In fact, the individual stars may be extremely distant from one other. A constellation is a pattern and a relationship – a meaning – we impose on the stars.

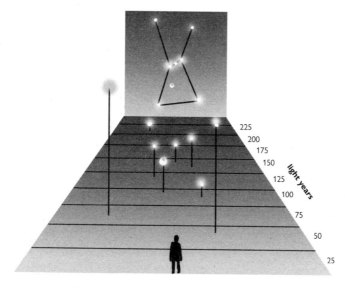

Figure 5: Distances to the stars in **60. Orion**

4.

OBSERVING: AN INTRODUCTION

WHERE TO START

Look up.

That's the first thing. We can often take the stars for granted as we go about our lives down on the ground. So look up and notice that the stars are there.

You'll see more stars by going outside to view them. Even without recognising any constellations, you can see that some stars are brighter than others. You might also discern different colours. Blue-white stars are relatively young; red stars are much older.

You can also differentiate the stars from artificial objects in the sky: the flashing lights on a plane, or the slow progress in a perfectly straight line of the International Space Station, high up in orbit.

What you can see will depend on the weather – a cloudless night can still suffer from atmospheric distortions (known as poor 'seeing', see page 28), which make the stars 'twinkle' and blur. Stars can also be obscured by the glare of a full or mostly full Moon.

You've already got this book, so this will be a great start to your observations. The star maps at the front and back of this book will help you to identify the patterns in the sky. Once you identify any of the 88 constellations, you can then look up the corresponding entry in Chapter 5. We hope that by learning a bit more about the constellations, you'll become more familiar with these patterns.

Some constellations are easier to see than others, so here are five to start with:

83. Ursa Major

In the northern hemisphere, this constellation is visible throughout the year. The tail section – also known as the Plough or the Big Dipper – is made up of seven very bright stars and is easily found.

84. Ursa Minor

The two brightest stars of **83. Ursa Major** (the outer edge of the Plough) can point you in the direction of Polaris, the North Star, which is in turn the tail of this smaller constellation.

9. Boötes

The arc of the Plough can guide you to the red giant Arcturus, the brightest star in this constellation. (You can remember this method easily: 'arc to Arcturus'.)

86. Virgo

Keep going from Arcturus and you can 'speed to Spica', a bright white star. This constellation is the only female character within the zodiac.

60. Orion

This constellation is visible during the winter of both hemispheres and it has a distinctive 'X' shape with a central belt of three stars. It includes two of the ten brightest stars in the night sky: Rigel and Betelgeuse. This constellation can also guide you to yet more stars (see **60. Orion** for more details).

Note that some constellations are only visible from the northern or southern hemisphere of Earth (not both). Some constellations are only visible for some of the year. We list these details in their entries in Chapter 5.

As well as constellations, there are many other things to observe in the night sky: planets, meteor showers, galaxies and nebulae. There's so much to see – so get looking.

NAKED-EYE OBSERVATIONS

You don't need specialist equipment to look at the stars. As we've seen, people of the ancient world studied and understood the patterns in the sky and the way they moved over the course of a year, all of which they did unaided. Telescopes weren't used in astronomy until the early 1600s. Some simple tips will improve what you can see with the naked eye.

It takes your eyes about 15 minutes to adapt to darkness. Your pupils need to fully dilate in order to take in more light, and when this happens you'll see the stars more clearly. But beware: if you then suddenly look at a bright light, it can be a little painful!

In addition, understanding the inner workings of the eye can help you to see more detail in fainter stars. Your eyes contain two kinds of receptor cells for detecting light: rods and cones. Rods are sensitive to even low levels of light, which is good for looking at stars in a dark night sky, but they're not good on detail and don't detect colour. By contrast, cones are sensitive to detail and colour, but they require a lot more light. Because of the way they're arranged in your eye, you can help focus starlight onto the rods by looking slightly to one side of the star in question. With some practice, this 'averted vision' technique can reveal stars 20 times fainter than you could see with direct vision.

Electric lights in urban areas – street lighting and people's homes – can obscure the stars, making fainter ones hard to see. You'll see more stars in places where there's less light pollution, such as in the countryside. But for a beginner, light pollution can be useful, as it makes only the brightest stars visible, which are the ones that make up the patterns in the constellations.

One more important thing: NEVER look directly at the Sun, as it is so bright that you can permanently damage your eyes. To look at the Sun safely, use a professional solar filter that blocks all visible wavelengths, plus infrared and ultraviolet, to a safe level.

BASIC KIT
A few inexpensive things will help your observations of the stars.

First, warm clothes – including gloves – are essential. Hats, scarves and blankets are all good. A hot water bottle, hand warmers and a Thermos of hot chocolate may also help – and can make you popular if you share them with other stargazers!

With a clipboard, paper, pencil and rubber you can make notes of your observations. A watch with a luminous dial will mean you can log the time of your observations.

An ordinary torch can be dazzling, which means that each time you use one you will have to wait another 15 minutes for your eyes to re-adapt to the dark. A red-light torch is a useful alternative so that you can see where you and your kit are, without affecting your night vision.

As well as the star maps contained in this book, you can use a planisphere: a simple, pivoted chart that can be adjusted to show the night sky, with labels, on the particular night you are observing. You can download a planisphere for free from https://in-the-sky.org/planisphere/index. php – a website with plenty of other advice and tips. A range of apps and devices are also available to guide you to particular stars and provide further information.

Binoculars or a handheld telescope will show you more detail than the naked eye. You may find it easiest to identify the constellations by naked eye and then use binoculars to home in on particular stars of interest.

We could write a whole separate book on the different kinds of telescope available, but here a few basic pointers.

The diameter of a telescope's objective mirror or lens, also known as the aperture, defines its light-gathering capacity. The bigger the aperture, the more light it collects, meaning fainter stars can be observed.

Resolution is the amount of detail that can then be seen. For example, a telescope with high resolution might enable you to see the two distinct bodies in a binary star.

Be wary of magnification, which spreads the same amount of light over a larger area without increasing resolution, so the object you see is larger but dimmer. Star clusters, galaxies and nebulae are often faint and cover a large area of the sky, so they are best viewed with low magnification. High magnification will help you see the detail of brighter objects that cover a small area of the sky, such as binary stars, but high magnification can also worsen 'seeing' (see 'Troubleshooting', below).

Some telescopes are fitted with GoTo technology, so they can automatically find and track particular objects in the sky. There are also a number of apps available (see page 30).

With a telescope stood securely on a tripod and focused on a particular star or constellation, you can often take serviceable photographs on your phone through the eyepiece of the telescope – but that's just the beginning for photography. Telescopic lenses are available for high-end cameras, and there are a range of apps and tutorials that will improve your photographs of the stars. Be warned: astrophotography can be addictive, and the kit can get expensive, but it's also enormously satisfying.

TROUBLESHOOTING

Various things can obscure our view of the stars.

The weather can prevent us from observing the stars at all, so it's a good idea to check the forecast before venturing outside. There's a famous old episode of *The Sky at Night* in which Patrick Moore gave a live demonstration of his own telescopes, and bad weather meant that he and the viewers couldn't see a thing!

Even on cloudless nights, a shimmering in the air called 'seeing' can make stars appear to twinkle and be less distinct. Seeing happens due to convection cells (pockets of warm, lower-density air) in Earth's atmosphere that distort the wavefronts of incoming light.

Two other things can make it harder to see fainter stars:

- Dust and water vapour in the atmosphere cause 'extinction' (absorption and scattering) of light, so that less light reaches us on Earth's surface.
- Light pollution is a particular problem in built-up areas such as cities. It causes the sky to glow, reducing the contrast between the sky and luminous celestial objects.

Since conditions can vary at different times, try observing the same stars and phenomena on different nights, and make notes so that you can compare what you see.

To overcome the difficulties of observation, astronomers use telescopes based well outside cities and at high altitudes – or even in space! Out-of-town observatories may host stargazing events for the public or offer slots to use high-altitude telescopes remotely. Some observatories offer ways for students to make use of their facilities, .but this doesn't mean that remote telescopes are reserved for the young. A friend of mine studied GCSE Astronomy at an adult night class, and for his coursework he had to program a remote-controlled telescope in Tenerife – by sending it instructions from his home in the UK – so that it photographed particular stars and phenomena.

Even if we can see a star or constellation clearly, visible light may provide us with only limited information; we can learn more by viewing stars via

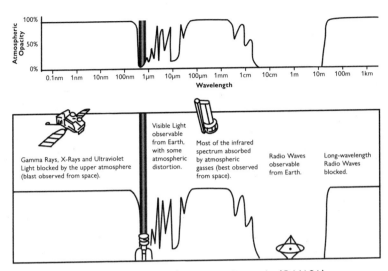

Figure 6: Atmospheric electromagnetic opacity (© NASA)

other wavelengths. However, while Earth's atmosphere is transparent to visible light and most radio waves, it is opaque to other forms of radiation, which makes it harder to view stars via these wavelengths:

- Radiation between 350 and 750nm can reach Earth's surface, so this is the so-called optical window that we can see.
- Radio waves greater than 30m are reflected by the ionosphere.
- Most infrared and microwave (that is, short-wavelength) radiation is absorbed by water and carbon dioxide. However, the atmosphere is transparent at some infrared (submillimetre) and microwave (millimetre) wavebands.
- Earth's atmosphere absorbs γ-radiation, X-rays and short-wavelength ultraviolet radiation.

To get a complete picture in astronomy, it is useful to collect data across the different wavelengths of the electromagnetic spectrum. Space-based telescopes outside Earth's atmosphere can detect energy that would not reach Earth's surface.

NAVIGATING THE HEAVENS
If we want to find a particular place on Earth, we use the grid system of latitude and longitude. We can navigate the night sky in much the same way, using the grid system of the celestial sphere.

This imaginary sphere has Earth at its centre, and onto the sphere are projected all the stars and other bodies of the night sky. The poles and equator of the celestial sphere extend outwards from the poles and equator on Earth.

Instead of latitude, the celestial sphere has declination (Dec.), the given position north or south of the celestial equator. This runs from the north celestial pole at +90°, through the celestial equator at 0°, to the south celestial pole at −90°.

Instead of longitude, the celestial sphere has right ascension (RA), measured in hours east along the celestial equator from a point called the vernal (or spring) equinox at 0° and up to 24 hours. Each hour of right ascension is the equivalent of 15° of longitude.

Traditionally, astronomers used declination and right ascension (and their own latitude and longitude) to calculate exactly when and where a particular star would appear in the sky above them. Today, you can enter the name of a star or constellation into an app or use GoTo technology to immediately direct you to the right part of the sky – but these technologies work by using this coordinate system.

Declination and right ascension apply wherever you are on Earth, but they are updated every 50 years to account for small changes in the positions of the stars relative to Earth. These 50-year periods are called epochs, and the last adjustments were made in the year 2000.

With the advent of smartphones, stargazing and identifying what you are observing has been made much easier through apps. These use your GPS location and the orientation of the phone to work out where your phone is pointing and therefore what should be visible in the field of view.

5.

THE CONSTELLATIONS

STAR LABELS

α	alpha	A
β	beta	B
γ	gamma	C
δ	delta	D
ε	epsilon	(short) E
ζ	zeta	Z
η	eta	(long) E
θ	theta	TH
ι	iota	I
κ	kappa	K
λ	lambda	L
μ	mu	M
ν	nu	N
ξ	xi	X
ο	omicron	(short) O
π	pi	P
ρ	rho	R
σ/ς	sigma	S
τ	tau	T
υ	upsilon	U
φ	phi	PH (F)
χ	chi	CH
ψ	psi	PS
ω	omega	(long) O

1. ANDROMEDA, THE 'CHAINED MAIDEN'

Pronounced: 'an-DROH-mih-duh'
Short: And
Brightest star: α Andromedae or Alpheratz (RA 0h 8m, Dec. +29°5')

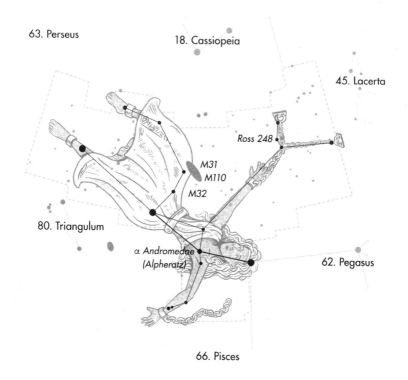

63. Perseus

18. Cassiopeia

45. Lacerta

Ross 248

M31
M110
M32

80. Triangulum

α Andromedae
(Alpheratz)

62. Pegasus

66. Pisces

*Andromeda consists of a total of 16 stars, with three stars brighter than magnitude 3 (for more on magnitudes, see page 120). These are slightly outshone by the easy-to-spot 'W' asterism of **18. Cassiopeia**, but the right side of the 'W' can be used to star hop · 5 from **14. Canis Major** to this constellation and the galaxy contained within its boundaries.*

WHEN AND WHERE TO OBSERVE

Andromeda is best viewed from the northern hemisphere in August and September, but is visible from June. In the southern hemisphere its appearance is much shorter, between October and December.

THE BRIGHTEST STARS

The brightest star in the constellation, α Andromedae, is generally thought to represent Andromeda's head (see 'Mythologies', below). Yet its Arabic name, Alpheratz, derives from *surrat al-faras*, usually translated as 'the naval of the steed', because it has also traditionally been seen as a part of the adjacent **62. Pegasus**. α Andromedae is located 97 light years from Earth and is, in fact, a binary star: a system of two stars orbiting a common centre of mass.

β Andromedae, also known as Mirach, is nearly as bright as Alpheratz and is located about 200 light years away from Earth.

γ Andromedae or Almach is Andromeda's foot. It is a multiple star system featuring a central giant 2,000 times more luminous than the Sun, orbited by a pair of white dwarfs. The third brightest star in the Andromeda constellation, Almach is about 350 light years away from Earth.

OTHER BODIES

As well as the stars that make up the constellation, other bodies lie within the confines of this constellation. Roughly halfway between the constellation's brightest star, α Andromedae, and the 'W' shape of **18. Cassiopeia** sits the Andromeda Galaxy (M31) (see page 22). This vast spiral galaxy, the largest neighbouring galaxy to our own Milky Way, is bright enough to be seen with the naked eye in dark skies, and with binoculars in built-up, light-polluted places. With large binoculars or a telescope, you can also see two more galaxies in the same part of the sky: M32 and M110.

METEOR SHOWERS

Between 26 October and 20 November, Earth's orbital path round the Sun takes it through debris left behind by Biela's Comet. Large pieces of rocky debris burn up in Earth's atmosphere, and seen from Earth's surface this meteor shower seems to radiate from Andromeda. As a result, the meteors are known as the Andromedids. The peak of the shower occurs around 8 November. This is one of the more minor meteor showers, and one can expect to observe around three meteors per hour. In recent years, the Andromedids have become less impressive due to the small movement of Earth's orbit around the Sun, meaning that our atmosphere interacts with less debris.

MYTHOLOGIES

According to ancient Greek and Roman mythology, Andromeda was an Ethiopian princess, daughter of **18. Cassiopeia** and **20. Cepheus**. Cassiopeia boasted that Andromeda was more beautiful than the Nereids, the legendarily good-looking sea nymphs. This claim caused serious offence, especially to the sea god Poseidon (or Neptune, in Roman versions of the story), who sent his pet, the monstrous **21. Cetus**, to lay waste to Ethiopia in response!

Cepheus was advised that the only way to save his kingdom was to sacrifice his daughter to Cetus. Andromeda was chained to a rock by the sea to await the monster, but at the last minute she was rescued by the young hero **63. Perseus**. In some versions of the story, Perseus flew to her rescue on the back of a winged horse, **62. Pegasus**. Perseus married Andromeda and they lived happily ever after. When they died, the gods placed them both among the stars to form the constellations.

In fact, the linking of this pattern of stars to a female character pre-dates the Greco-Roman tradition. Long before that, the ancient Babylonians associated some of the same stars with a fertility goddess, Anunitum. The connection to Andromeda dates back at least to the second century CE, when astronomer Claudius Ptolemy (c. 100–170 CE) included it as one of the 48 constellations in his book *Almagest*. We know that Ptolemy drew on earlier sources, so the connection with Andromeda may date much further back in time.

INTERESTING FACTS

Launched from Earth on 20 August 1977, the Voyager 2 space probe is heading in the direction of Andromeda. However, it will take 40,000 years for the probe to reach the nearest of the constellation's stars: at 10.3 light years away, red dwarf Ross 248 is too faint to be seen by the naked eye from Earth. Sadly, Voyager 2 will see little of it either, as the probe is expected to get no closer than 1.7 light years (16 trillion kilometres) from the star. Besides, the power for the probe's instruments is not expected to last much beyond the year 2025. (Another space probe, Voyager 1, is heading in the direction of **11. Camelopardalis**.)

Our Milky Way galaxy is due to collide with the Andromeda Galaxy in around 4.5 billion years. Although this sounds catastrophic, because there is so much space between adjacent stars there are likely to be few actual collisions.

2. ANTLIA, THE '[AIR] PUMP'

Pronounced: 'ANT-lee-uh'
Short: Ant
Brightest star: α Antliae (RA 10h 27m, Dec. −31°4')

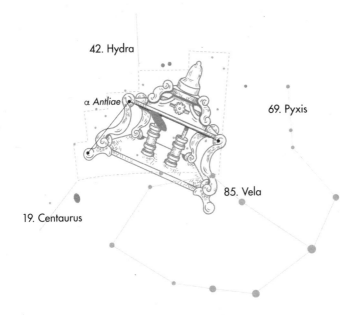

42. Hydra

α Antliae

69. Pyxis

85. Vela

19. Centaurus

*After the romantic tales of **1. Andromeda** and the challenges she faced, having a constellation called the Air Pump seems a little down to earth. Antlia is a small constellation consisting of four faint stars. It is one of 14 constellations first identified and named by the French astronomer Nicolas-Louis de Lacaille (1713–62). Whereas many older constellations are named after characters and creatures from ancient mythology, Lacaille named his after the latest scientific tools of the day. It's a little like if an astronomer now were to announce a constellation called 'the 3D printer' or 'the Spectroscope'.*

WHEN AND WHERE TO OBSERVE
This constellation is visible in southern skies from January to March. It can be seen between latitudes of +45 and −90°.

THE BRIGHTEST STARS
The brightest star, α Antliae, is a K-type orange giant, more than twice the mass and 40 times the diameter of our Sun.

OTHER BODIES

As this constellation is quite small, there are not many other bodies in the area. Planet HD93083 b was discovered in 2005 and sits around 94 light years away from our Solar System. It seems to be in the habitable zone around its star, the right distance to allow water, and even life, to exist. But we think it's also a gas planet, so life there might be very unusual.

METEOR SHOWERS

There are no significant meteor showers associated with this constellation.

MYTHOLOGIES

Being a bit of a fill-in constellation, there does not seem to be much mythology around these stars. However, Antlia can be observed from China, and two of its stars were included in two Chinese constellations, including Dong'ou, which represents an area in southern China.

INTERESTING FACTS

Since Lacaille named so many constellations, it's worth knowing who he was. Having studied philosophy and theology, Lacaille became an *abbé*, a low-ranking member of the clergy in France's Catholic Church. He took a job as a surveyor and soon made a name for himself by accurately measuring a long stretch of the French coast and by remeasuring the 'French meridian'.

A meridian is an imaginary line that runs straight from the North Pole to the South Pole (well, 'straight' given that the line must arc over Earth's curved surface!). We also know meridians as the lines of longitude in the grid system we use to map Earth. Longitude is measured in degrees east or west of the prime meridian, which is the 'zero' line that passes through the Royal Observatory in Greenwich, England. The grid we use to map the night sky and stars is an extension of this system.

Lacaille's French meridian is the line of longitude that passes through the Paris Observatory in France, 2°20' east of the prime meridian. It took two years of painstaking work as a surveyor before Lacaille published his findings, but the result was met with acclaim. His revised meridian meant that better, more accurate maps could be made of France, and it also meant that astronomers could calculate a more accurate figure for the size of Earth, and in doing so could better judge the size and distance of other celestial objects, such as the planets of the Solar System.

But Lacaille didn't stop there. He proposed that the distance of the planets could be calculated even more accurately by observing them from two separate points on Earth's surface and then comparing the results using trigonometry. As a result, while a colleague worked in France, Lacaille established a new observatory at Table Bay in what is now South Africa and took careful observations every night for two years. In the course of this work, he recorded observations of 9,766 stars visible from the southern hemisphere, and he grouped some of these into 14 new constellations.

Lacaille's astronomy was informed by his philosophy. He was influenced by Enlightenment ideas then popular in Europe: very basically, that by using evidence and reason we can better know the universe around us and so live freer, happier lives. Ancient myths were a way to explain the world and our place in it; for Lacaille, scientific tools didn't just explain the world but could lead to progress. His observations and the names he coined for the new constellations were all part of an effort to build a better future.

He initially named this particular constellation *la Machine Pneumatique*, meaning a machine operated by air or gas under pressure, and he seems to have had in mind a single-cylinder vacuum pump used in scientific experiments by French physicist Denis Papin (1647–1713). Some later publications depicted the constellation as the more powerful double-cylinder pump. In the charts published a year after Lacaille's death, the name was changed to Latin, *Antlia pneumatica*, since shortened to Antlia.

Antlia can be observed below latitude +49°– that is, southward from Paris – but it's a relatively faint constellation.

3. APUS, THE 'BIRD OF PARADISE'

Pronounced: 'APE-us'
Short: Aps
Brightest star: α Apodis (RA 14h 47m, Dec. −79°2')

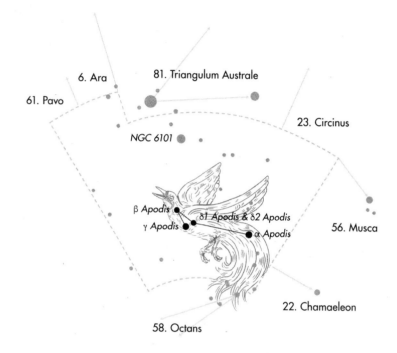

Apus is a small constellation visible in the southern hemisphere. It consists of four stars, but many others sit within its boundaries.

WHEN AND WHERE TO OBSERVE

This constellation sits in the southern hemisphere close to the South Pole. It is visible for most of the year in the southern hemisphere but is most visible during July.

THE BRIGHTEST STARS

Brightest star α Apodis has an apparent magnitude of 3.825 and is approximately 410 light years away. It is classified as a K-type giant. Both α Apodis and β Apodis are orange giants. Though they look similar, α Apodis is actually much larger, but it is also more than twice as distant from Earth. γ Apodis is a warmer yellow star, and it is just possible to see with the naked eye that δ Apodis is a double star – two stars that look very close when viewed from Earth (for more information on double stars, see **14. Canis Major**). In fact, δ^1 Apodis is a much more distant red giant, and δ^2 Apodis is more orange; the difference in colour is more discernible through binoculars.

OTHER BODIES

Two stars within Apus are thought to have exoplanets in orbit around them. There are at least two globular clusters associated with this constellation.. A small telescope will show one of these, NGC 6101, as a hazy spot. In fact, this is a globular cluster (see 'Star Formation', page 8) thought to contain an unusually high number of black holes (see 'Life Cycle of Stars', page 9).

METEOR SHOWERS

There are no significant meteor showers associated with this constellation.

MYTHOLOGIES

As this constellation contains stars that are relatively dim, there are no known mythologies associated with this constellation.

INTERESTING FACTS

In September 1519, the Portuguese explorer Ferdinand Magellan and a fleet of five ships set off from Seville in Spain on an extraordinary voyage. Their destination was the Maluku Islands in eastern Indonesia, known as the 'Spice Islands', as they were at the time the only source of valuable cloves, mace and nutmeg.

What made Magellan's trip so extraordinary was that he didn't follow the trade routes used since ancient times but headed westwards from Seville, crossing the Atlantic and then the Pacific. Magellan died before the fleet reached the Maluku Islands, but the expedition went on without him and was a success. The surviving crew returned home by continuing west, arriving back in Seville in September 1522 and becoming the first ever people to complete a voyage right around the world!

As well as precious spice, the crew returned home with lots of other treasures from their expedition, including examples of the dazzlingly plumed birds of paradise largely found in and around the islands of Papua New Guinea. These weren't living specimens – that would hardly have been practical on such a long voyage – and it seems that their legs were removed because they were thought to distract from the beautiful plumage. As a result, some Europeans thought birds of paradise didn't have legs, and the species was named *Apus*, from the Greek meaning 'without feet'.

Less than a century later, the astronomer Petrus Plancius applied the name Apus to a group of relatively faint stars near the celestial south pole that cannot be seen from Earth's northern hemisphere. (For more about Plancius and the 15 constellations he identified, see **30. Crux**.)

4. AQUARIUS, THE 'WATER BEARER'

Pronounced: 'ah-KWAIR-ee-us'
Short: Aqr
Brightest star: β Aquarii or Sadalsuud (RA 21h 31m, Dec. −5°34')

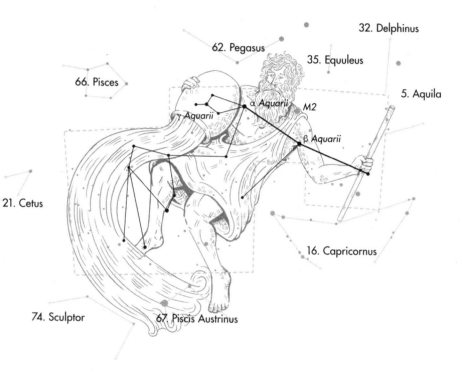

Although this pattern of stars has been recognised since ancient times, Aquarius is one of the harder constellations to spot as it does not contain any bright stars.

WHEN AND WHERE TO OBSERVE

One of the constellations of the zodiac, Aquarius can be seen in the northern and southern hemispheres. Aquarius lies on the ecliptic, which is different to the celestial equator: the celestial equator is the projection of the terrestrial equator into space; the ecliptic is the apparent path that the Sun takes across the sky.

In the northern hemisphere, Aquarius is best seen in the autumn evening sky. It lies east of another zodiacal constellation, **16. Capricornus**. You can use the Great Square of **62. Pegasus** to guide you towards Aquarius.

THE BRIGHTEST STARS

The two brightest stars in Aquarius are yellow supergiants. β Aquarii is very slightly brighter than α Aquarii and some 2,200 times more luminous than our Sun. M2 appears as a fuzzy spot through binoculars, but a telescope shows it to be a good example of a globular star cluster.

Although 14 stars make up the pattern of the water bearer in Aquarius, scans of this area have detected around 2,100 stars. On a very clear night up to 188 of these are visible with the naked eye.

OTHER BODIES

A number of deep-space objects lie within the boundaries of Aquarius. We know of 40 exoplanets orbiting stars within this constellation. As well as these, there are two globular clusters and an asterism of four stars.

METEOR SHOWERS

Three major meteor showers appear to originate in the constellation Aquarius: the Eta Aquariids, the Delta Aquariids and the Iota Aquariids. The names indicate the star in the constellation from which these 'shooting stars' appear to derive. The Eta Aquariids are the strongest meteor shower radiating from Aquarius; the shower peaks between 5 and 6 May with approximately 35–50 meteors per hour. The Delta Aquariids occur between 12 July and 23 August, peaking around 30 July each year. The Iota Aquariids occur between 11 August and 10 September, peaking around 25 August.

MYTHOLOGIES

The name Aquarius comes from the Latin for 'water bearer', and it's one of the longest-recognised patterns of stars. The Babylonian text MUL. APIN, compiled in about 1000 BCE from earlier sources, refers to the constellation as 'the Great One'; it was thought to show the water god Ea with a huge, overflowing vase, which the Babylonians considered to be the source of regular, catastrophic floods.

The ancient Egyptians also associated this figure of a water bearer with their own water god – called Hapi – and with flooding. However, they took a more positive view of this figure because regular flooding of the Nile was essential for Egyptian prosperity.

Greek and Hindu traditions also saw in these stars a vase from which streamed water, suggesting that similar ideas may have spread through different cultures. It seems Aquarius also influenced the patterns seen in the stars around it: neighbouring constellations include a whale (**21. Cetus**), a dolphin (**32. Delphinus**) and two fish (**66. Pisces** and **67. Piscis Austrinus**). Some also linked this water-bearing figure to the legendary Trojan prince Ganymede.

INTERESTING FACTS

TRAPPIST-1 is an ultracool dwarf star in Aquarius, thought to have a surface temperature of just 2,556K. We know at least seven rocky planets orbit this small star, and three of these may sit within the 'habitable zone', so might have liquid water on them (see '"Habitable" or "Goldilocks" Zones', page 61).

5. AQUILA, THE 'EAGLE'

Pronounced: 'ACK-will-uh'
Short: Aql
Brightest star: α Aquilae or Altair (RA 19h 50m, Dec. +8°52')

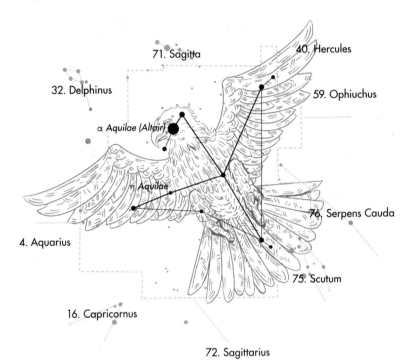

This eagle-shaped pattern of stars was recognised by the astronomers of ancient Babylon and Sumer in what is now Iraq. It's thought that this interpretation was passed down to the Greeks. It is the 22nd largest constellation in the sky and contains the asterism known as the Summer Triangle.

WHEN AND WHERE TO OBSERVE
Aquila is best seen in the northern hemisphere during the summer. It is located in the northern sky, near the celestial equator, and so can be seen at latitudes between +90° and –70°. Aquila can be seen in the southern hemisphere between July and November, where it sits in the northern Milky Way.

THE BRIGHTEST STARS
α Aquilae or Altair has an absolute magnitude of 2.22. It sits approximately 16.7 light years away from Earth and is the 12th brightest star in the night sky.

Altair is a white, A-type star that spins extremely rapidly. Measured at the equator, the star takes about nine hours to complete a rotation, whereas our Sun takes about 25 days. This high velocity makes the star bulge outwards – a process called oblation. As a result, the diameter at the equator is 20% greater than the diameter measured pole to pole.

OTHER BODIES

Of the many stars contained in the area of this constellation, nine are thought to have exoplanets around them. There are planetary nebulae contained within this constellation: the Glowing Eye Nebula (NGC 6751) and the Phantom Streak Nebula (NGC 6741). There are also a number of open and globular clusters.

METEOR SHOWERS

Two meteor showers appear to radiate from this constellation. The Northern June Aquilids were discovered quite recently, in 1976, and are active from around 26 June to 22 July, peaking around 15 July. The Epsilon Aquilids occur between 4 May and 27 May, peaking close to 17 May.

MYTHOLOGIES

The Greek name Aquila, meaning 'eagle', dates to at least as long ago as the writings of mathematician and astronomer Eudoxus of Knidos (c. 400– c. 350 BCE) in what is now Turkey. Although his works are long since lost, fragments are quoted in later surviving texts.

Some writers linked Aquila to particular eagles in Greek mythology. One candidate was the eagle that supposedly carried thunderbolts for Zeus, father of the Greek gods. In another story, Zeus transformed himself into an eagle to snatch up Trojan prince Ganymede and fly him away to spend eternity serving drinks to the gods. In this version, Ganymede was also immortalised in the stars, as the constellation **4. Aquarius**, the 'Water Bearer'.

The bright star α Aquilae is also known as Altair, a shortened version of an Arabic phrase meaning the 'flying eagle'. In Australia, some aboriginal cultures also link this star to eagles: for the Kulin people in the area around Melbourne and for the Wotjobaluk people of western Victoria, this eagle is creator-god Bunjil; for the Wardaman people of the Northern Territory, it is Bulyan, an eagle that watches over **26. Corona Australis**.

INTERESTING FACTS

The classic science-fiction film *Forbidden Planet* (1956) – which was a huge influence on *Star Trek* – is set on Altair IV, presumably the fourth planet orbiting α Aquilae. In the film, the planet is Earth-like, with a breathable atmosphere rich in oxygen, and seems ideal for people to live on – that is, until we learn the terrible fate that befell the planet's native civilisation.

6. ARA, THE 'ALTAR'

Pronounced: 'AR-uh'
Short: Ara
Brightest star: β Arae (RA 17h 25m, Dec. −55°31')

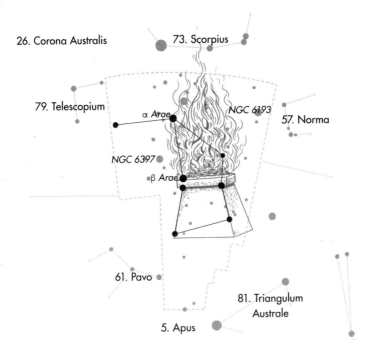

The ancient Greeks saw in this pattern of stars an altar with smoke 'rising' from a ritual offering. From the northern hemisphere, Ara appears low in the sky so the smoke doesn't rise 'up' but falls away southward, past the horizon. It's as if the offering is being made to something beyond Earth or below the ground — where, it was thought, the gods resided.

WHEN AND WHERE TO OBSERVE
This small constellation in the southern sky is visible between latitudes of +25° and −90° and is best observed in July.

THE BRIGHTEST STARS
In his 1603 star atlas *Uranometria*, German astronomer Johann Bayer (1572–1625) identified eight stars in Ara and listed them in order of brightness from α to θ (see page 21 for more about Bayer designations). In fact, orange supergiant β Arae is very slightly brighter than blue-white α Arae.

OTHER BODIES

NGC 6193 can be seen with the naked eye. It's an open cluster with three hot, bright blue-white O-class stars in orbit around one another, plus another O-class star that may be (we're not sure) two stars in orbit around each other, and at least 20 slightly cooler B-type stars – a sizeable 'family' of stars that all formed at about the same time from one vast cloud of gas.

At about 7,800 light years from Earth, NGC 6397 is one of the two closest globular clusters to us (the other, M4, is in **73. Scorpius**). However, you'll need good binoculars or a telescope to see it.

METEOR SHOWERS

There are no significant meteor showers associated with this constellation.

MYTHOLOGIES

Some ancient people saw in this constellation the altar at which the Olympian gods led by Zeus made offerings together to cement their alliance before waging a ten-year war against the Titans – a foundational moment in Greek mythology. Others thought it was the altar belonging to the wise centaur Chiron (see **19. Centaurus**).

Different cultures could apply their own meanings, even where they saw the same pattern in the stars. Outside the Greek tradition, Ara has been identified as the altar constructed by Noah after he and his family survived the Great Flood, and also as the altar from Solomon's famous temple in Jerusalem.

INTERESTING FACTS

Seven of the star systems in Ara are known to have exoplanets. The star μ Arae, or Cervantes, is thought to be very like our Sun and is orbited by four known exoplanets: Quijote, Dulcinea, Rocinante and Sancho. The names are taken from characters in the famous novel *The Ingenious Gentleman Don Quixote of La Mancha* by the Spanish writer Miguel de Cervantes (1547–1616).

7. ARIES, THE 'RAM'

Pronounced: 'AIR-eez'
Short: Ari
Brightest star: α Arietis or Hamal (RA 2h 7m, Dec. +23°27')

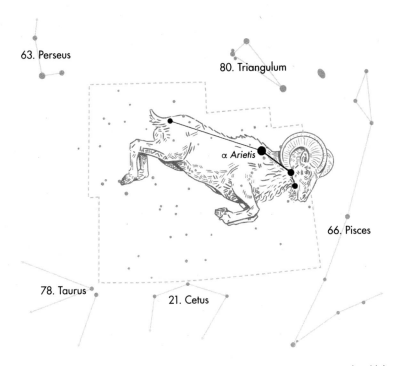

This is one of the smaller constellations of the astrological zodiac: the 11th smallest of the 12. The pattern that makes up the 'Ram' contains about nine stars, but only four are particularly bright, with magnitudes below 4.

WHEN AND WHERE TO OBSERVE

Aries can be seen in both the northern and southern hemispheres. It can be seen during spring and midsummer in the southern hemisphere. In the northern hemisphere it can be seen during autumn and winter.

THE BRIGHTEST STARS

The brightest star, α Arietis, is also known as Hamal, a shortened version of the Arabic phrase for 'head of the ram'. Precession (see 'Interesting facts', below) means that the star's declination is now almost the same as latitude 23°26', known as the Tropic of Cancer (see 'Tropic of Cancer', page 73), and navigators can calculate their position relative to that tropic from the height of this star in the sky.

OTHER BODIES

There are no Messier objects (see page 8) within the constellation, but there are a number of star clusters and galaxies. There are currently thought to be 17 exoplanets orbiting stars within this constellation.

METEOR SHOWERS

The Arietids are one of the few meteor showers that can be seen during the day. They occur between 22 May and 7 July, peaking on 7 June. The shower was discovered in 1947 by the telescopes at Jodrell Bank in the UK.

MYTHOLOGIES

The idea that this pattern of stars represents a ram seems to have developed over time. The Babylonians initially saw it as an 'agrarian', or farm worker, perhaps because the constellation appeared in spring when crops were being sown. Later, things got more specific and the same pattern was identified as the shepherd god Dumuzi. In Egypt, the constellation was identified as the ram-headed god Amon-Ra. To the Greeks it was the fabled flying ram with a golden fleece. It's surely no coincidence that three neighbouring cultures all saw a ram in the same stars, suggesting that ideas spread – we don't know whether purposefully or not – and were then adapted to local mythology.

INTERESTING FACTS

Earth spins on its axis while also orbiting the Sun. When the surface of Earth is exposed to the Sun's heat and light, it is 'day' for that part of the planet, while for the part in shadow it is 'night'. But day length varies during the course of an orbit (or a year): days are longer than nights in summer and shorter than nights in winter. This is because Earth spins on an axis angled 23.4° to the plane of its path round the Sun.

Twice a year, when the Sun is directly over Earth's equator, day and night are of equal length. These are the equinoxes, meaning 'equal nights'. In the northern hemisphere, the spring or 'vernal' equinox occurs around 21 March and the autumn equinox around 21 September. In the southern hemisphere, it's the other way round.

In the northern hemisphere, the vernal equinox has long been associated with spring and new life, and for many ancient people it marked the beginning of a new year. Between about 2000 BCE and 100 BCE, at vernal equinox the Sun appeared to be passing through the pattern of stars now called Aries, so this was seen as the 'first' sign of the zodiac in the annual cycle. Aries is still usually listed first in newspaper horoscopes.

However, the gravity of the Sun and Moon cause a slight wobble in Earth's axis of rotation, a shift of about 1° every 70 years, in a regular cycle lasting 25,772 years and known as precession. The effect of this is that, since ancient times, the vernal equinox has shifted and is now in **66. Pisces**. In about 700 years or so, it will pass into **4. Aquarius**.

8. AURIGA, THE 'CHARIOTEER'

Pronounced: 'AW-rye-gah'
Short: Aur
Brightest star: α Aurigae or Capella (RA 5h 16m, Dec. +45°59')

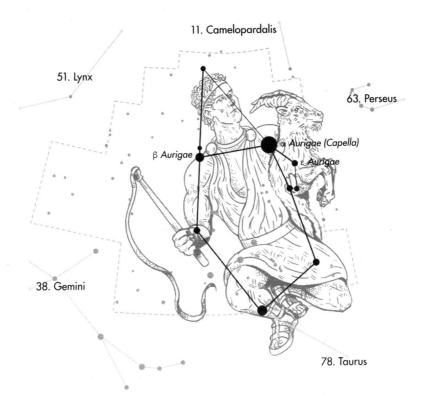

The jewel in this constellation is the bright star Capella. It is the sixth brightest star in the night sky and third brightest in the northern hemisphere. To the naked eye it appears to be one star, but it is actually a quadruple star system made up of two binary pairs.

WHEN AND WHERE TO OBSERVE
Auriga is visible in the evenings in the northern hemisphere from late autumn and through winter. It is visible at latitudes +90 to −40°.

THE BRIGHTEST STARS
We now know that α Aurigae (Capella) is a double binary star (see **14. Canis Major**) and the two large stars are closer to each other than

Venus is to the Sun. The other binary is made up of two fainter dwarf stars. The system also seems to be a bright source of X-rays due to emissions from the superheated gas/plasma surrounding the binary stars.

β Aurigae is also a binary star, comprising two blue-white A-types. ε Aurigae is another binary star. Seen from Earth, the brighter white supergiant is eclipsed by its less bright counterpart every 27 years, its apparent magnitude dropping from 3 to 3.8 in the process.

OTHER BODIES

A number of star-forming open clusters sit within this constellation, but you'll need binoculars to observe even the brightest of these.

METEOR SHOWERS

The brightest meteor shower associated with this constellation is the Aurigids. The meteors originate from Comet Kiess, discovered by Carl Kiess in 1911. This is an example of a 'long-period' comet, which takes around 2,000 years to orbit the Sun. Even though the comet's orbital period is long, the trail of debris left behind gives us the Aurigids every year, peaking around 1 September.

MYTHOLOGIES

Being so bright, Capella has featured in the constellations of many cultures. In Hindu mythology, it was thought to be the heart of creator-god Brahma. In some aboriginal mythology, Capella was Purra, the kangaroo, being chased across the sky by Yurree and Wanjel (twins Castor and Pollux in **38. Gemini**).

The ancient Greeks saw in this pattern of stars one of three legendary charioteers. One, Erichthonius, had a disability that made it difficult to walk. It's said he took inspiration from the belief that the Sun moved swiftly across the sky in a chariot, and so invented the quadriga – an especially fast chariot pulled by four horses side by side. According to legend, the quadriga gave Erichthonius and his followers a crucial advantage over their enemies in battle, and he became king of Athens. Zeus rewarded his heroism by placing him in the stars.

Alternatively, the charioteer is Myrtilus, who killed King Oenamaus by sabotaging his chariot before a race, swapping the strong bronze clasps that held it together with weak ones made of beeswax. In both these stories, there's a sense that the chariot is so fast and dangerous that it could change the balance of power.

The constellation was also seen as Hippolytus (son of the more famous Theseus, from the legend of the Minotaur), who died in a chariot crash but was brought back to life by brilliant doctor Asclepius. Whatever the case, many early depictions of the constellation only show the braced standing position of the charioteer, not the chariot itself; the focus is on the skill of the human rider.

We can contrast this with the way many of the same stars were seen in Chinese astronomy: as five imperial chariots in the constellation 五車 (Wuche), and as pillars to which horses could be tethered in 柱 (Zhù).

INTERESTING FACTS

Some Western depictions of Auriga show the charioteer with a goat on his shoulder, and sometimes with smaller goats under his arm. This is because α Aurigae was long known as the 'goat star'. Around 1000 BCE, it was known as such by the Babylonians, long before the Greeks linked it to Amalthea, the legendary goat whose milk nurtured the young Zeus. The traditional name for the star, Capella, derives from 'little goat'.

In fact, this and other nearby stars were seen as their own constellation – the young goats, or kids – by astronomer Cleostratus in about 400 BCE. Ptolemy's *Almagest*, written in about 150 CE, is the earliest surviving source to combine the goats with the charioteer.

9. BOÖTES, THE 'HERDSMAN'

Pronounced: 'bo-OH-teez'
Short: Boo
Brightest star: α Boötis or Arcturus (RA 14h 15m, Dec. +19°10')

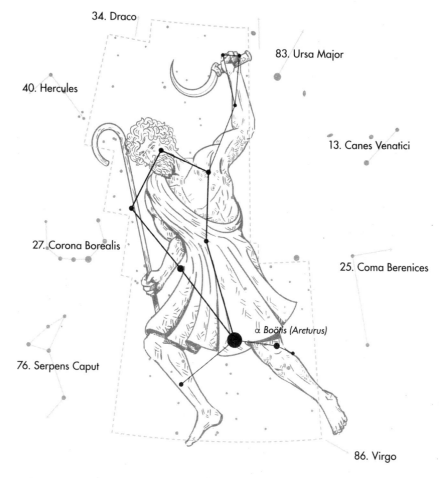

Boötes is one of the largest constellations and contains the third brightest star in the night sky.

WHEN AND WHERE TO OBSERVE

In the northern hemisphere, Boötes is best observed during spring and summer evenings, though its extended arm is visible throughout the year. The constellation is visible in the southern hemisphere during autumn and winter.

THE BRIGHTEST STARS

α Boötis has been known as Arcturus (from the Greek for 'guardian of the bear') since the time of Homer. It's a red giant – about the same mass as the Sun but about 25 times its size – that looks distinctly yellow-orange to the naked eye. Arcturus is a little less than 37 light years from our Sun and is thought to be around 7 billion years old.

Arcturus is very easily found by first locating the curved 'handle' of the Plough in **83. Ursa Major**: you then follow the arc to Arcturus. By continuing straight on, you can also 'speed to Spica', the brightest star in **86. Virgo**.

OTHER BODIES

Unlike many other constellations, Boötes is famous for having very few other bodies visible within its boundaries. The Boötes Void is a volume of space some 250 million light years in diameter where very little can be seen. In recent years, up to 60 faint galaxies have been discovered in the region.

METEOR SHOWERS

The Quadrantid meteor shower occurs in this constellation. Although this has as many as 100–130 meteors an hour, it is not one of the more popular showers as its 'shooting stars' are not very bright and the shower can be short-lived, some years only lasting an hour or so.

MYTHOLOGIES

The name Boötes derives from βοῦς, the Greek for 'cow', and refers to a human herdsman or ox driver. The constellation is cited in Homer's famous epic poem, the *Odyssey*, thought to have been written in about 750 BCE. Yet we're not sure exactly who this herdsman was or what he was herding, as the ancient Greeks linked Boötes to several different figures from myth.

One was Philomelus, a demigod associated with farming, who was immortalised in the night sky as reward for inventing the plough, which was of such benefit to humanity. Yet, perhaps strangely, the Greeks don't seem to have recognised his invention in the nearby group of stars we now know as the Plough (in **83. Ursa Major**). To them, those stars made up the oxcart Boötes is herding.

This interpretation may have been inherited from other, earlier cultures. To the Babylonians, the same stars comprising Boötes depicted Enlil, leader of the gods, who was thought to have invented another key agricultural tool, the mattock – which has a blade like an axe, but at a right angle to the handle. Perhaps the Greeks were also influenced by the ancient Egyptians, who saw in the Plough the foreleg of an ox.

In another reading of Boötes recorded by Aratus in the 200s BCE, he is Arctophylax, the 'Bear Keeper', herding the bears **83. Ursa Major** and **84. Ursa Minor** round the celestial north pole. In some depictions of this story, Boötes is aided by the hunting dogs **13. Canes Venatici**. Around the same time as Aratus, others associated Boötes with Arcas, a son of Zeus, who hunted a bear without realising that it was really his mother Callisto transformed.

Roman scholar Gaius Julius Hyginus (c. 64 BCE–17 CE) had yet another story. According to him, Icarius was a winemaker from Athens who was murdered by shepherds who thought his wine had poisoned them. In their grief, Icarius' daughter Erigone and his faithful dog both died soon afterwards, and Zeus was so touched by the sad story that he placed them all in the night sky: Icarius as Boötes, Erigone as **86. Virgo** and the dog as either **14. Canis Major** or **15. Canis Minor.**

INTERESTING FACTS

In Hawaiian culture, the bright star Arcturus was thought to be called Hokule'a, which roughly translated means 'star of gladness'. The star itself was important for navigation.

10. CAELUM, THE 'ENGRAVING'

Pronounced: 'SEE-lum'
Short: Cae
Brightest star: α Caeli or (RA 4h 40m, Dec. −41°51')

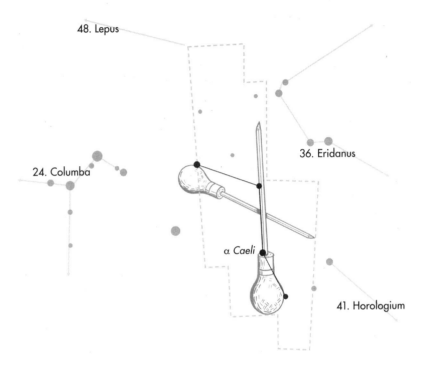

*This small constellation of relatively faint stars is one of the most difficult to see and is another of Lacaille's 14 fill-in constellations (see **2. Antlia**).*

WHEN AND WHERE TO OBSERVE
Our tip for finding this small, faint constellation in the skies of the southern hemisphere is to first locate the more readily visible Canopus, the brightest star in **17. Carina**, and then pan eastwards.

THE BRIGHTEST STARS
Even the constellation's brightest star, α Caeli, has an apparent magnitude of only 4.5 so is too faint to be seen from light-polluted cities. It's a binary star comprising an F-type star a little bigger than the Sun and a cooler, dimmer red dwarf.

OTHER BODIES

As this constellation is relatively small and out of the plane of our Milky Way galaxy, not many deep-space objects have been observed in the confines of this constellation. One of the only noteworthy objects seen in the area is known as HE0450-2958. This is a 'quasar': a highly luminous body millions to billions of times more massive than our Sun. What makes quasars so bright is that they sit in galactic centres and are powered by the supermassive black holes found at the nuclei of galaxies. As matter falls into the supermassive black hole, it heats up due to friction and this generates huge amounts of radiation. What makes HE0450-2958 so interesting is that the galaxy that it should sit in has not, as yet, been observed.

METEOR SHOWERS

There are no significant meteor showers associated with this constellation.

MYTHOLOGIES

As Caelum is made up of faint stars, it has mostly gone unnoticed by the cultures of the world, so little to no mythology can be found about the stars here.

INTERESTING FACTS

Lacaille originally named this constellation Burin, the word for an engraver's cutting tool. This was later Latinised as Caela Sculptoris, the 'sculptor's tool', and then shortened to Caelum.

11. CAMELOPARDALIS, THE 'GIRAFFE'

Pronounced: 'ka-MEL-oh-PAR-duh-liss'
Short: Cam
Brightest star: β Camelopardalis (RA 5h 3m, Dec. +60°26')

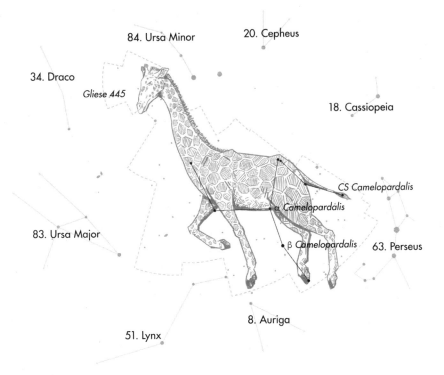

This pattern of stars is one of 15 constellations first recognised by the Dutch astronomer Petrus Plancius (see **30. Crux**). In about 1613, Plancius named it Camelopardalis, the Latin word for 'giraffe', derived from the Greek for 'spotted camel'.

WHEN AND WHERE TO OBSERVE

This is a large constellation close to the north celestial pole, so it is visible throughout the year in the northern hemisphere, but its lack of bright stars means it can be difficult to spot.

THE BRIGHTEST STARS

Camelopardalis is a relatively faint constellation. With an apparent magnitude of 4.03, the brightest star, β Camelopardalis, is a yellow, G-class supergiant 58 times the diameter of the Sun.

The binary star CS Camelopardalis is the next brightest – a blue-white B-type supergiant with a variable magnitude of between 4.19 and 4.23.

OTHER BODIES
Kemble's Cascade is a group of more than 20 stars that, seen from Earth, seem to form an almost straight line. In 1980, this asterism was named after Franciscan friar and amateur astronomer Father Lucian Kemble (1922–99), who had enthused about seeing it.

METEOR SHOWERS
The Camelopardalids are visible on 23 and 24 May each year in the northern hemisphere, the result of debris from comet 209P/LINEAR.

MYTHOLOGIES
It's sometimes claimed that Camelopardalis means 'camel-leopard', and that the ancient Greeks thought a giraffe was a mixture of these two animals. In fact, the 'pardalis' part means 'spotted', so they saw the giraffe as a 'spotted camel', while the leopard was a 'spotted lion' (see **46. Leo**.)

INTERESTING FACTS
Launched in 1977, space probe Voyager 1 is moving in the direction of this constellation.

12. CANCER, THE 'CRAB'

Pronounced: 'KAN-ser'
Short: Cnc
Brightest star: β Cancri or Altarf/Tarf (RA 8h 16m, Dec. +9°11')

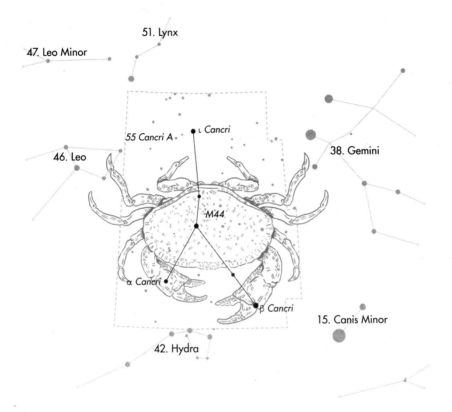

The name Cancer may be familiar because this is one of the 12 constellations of the zodiac; the constellation itself is of medium size and does not contain any particularly bright stars. The earliest known reference to this constellation is in Almagest, written in about 150 CE by Greek-Egyptian astronomer Claudius Ptolemy (c. 100–170 CE). He labelled it Karkinos, the Greek for 'crab'. Some say Ptolemy invented this and 47 other constellations, but it's likely that at least some of them were drawn from earlier sources now lost.

WHEN AND WHERE TO OBSERVE

This constellation is visible in the evening sky of the northern hemisphere in early spring, and in the southern hemisphere in early autumn. Since the constellation is not very bright, finding it can be challenging, but a good tip is first to locate the crescent mane of **46. Leo** and twin bright stars Castor and Pollux in **38. Gemini**. Cancer lies halfway between them.

THE BRIGHTEST STARS

β Cancri is the brightest star in the constellation, an orange K-type giant also known as Altarf or Tarf from the Arabic for the 'end [of the crab]'. The star is actually 870 times more luminous than our Sun. It's also a binary star (see **14. Canis Major**), though it is 2,600 times more distant from its companion star – a faint M-class red giant – than Earth is from the Sun. At that distance, the two stars complete an orbit of one another every 76,000 years!

In 2014, scientists reported evidence of a planet in orbit round β Cancri almost eight times as massive as Jupiter (the largest planet in the Solar System).

55 Cancri A is thought to have five planets in orbit around it: one super-Earth and four gas giants, one of them in the habitable zone (see '"Habitable" or "Goldilocks" Zones', page 61).

OTHER BODIES

M44 is easily visible to the naked eye in dark locations and is popular with amateur astronomers. Also known as Praesepe or the 'Beehive', it's an open cluster of more than 200 stars, which can be more readily seen through binoculars or a telescope. In fact, this cluster was one of the first celestial objects ever to be studied by telescope: Galileo Galilei constructed his first telescope in 1609, and the following year made sketches of 'Nebulosa' or cloudy Praesepe and also of **60. Orion**. We now know that, at some 610 light years away, M44 is one of the closest such clusters to Earth.

METEOR SHOWERS

The Delta Cancrids peak on January 17 each year but are a relatively weak meteor shower, with fewer than five meteors per hour.

MYTHOLOGIES

This pattern of stars was associated with other, similar creatures long before Ptolemy likened it to a crab. To the Babylonians it was MUL. AL.LUL, which can mean either 'crab' or a type of turtle; surviving depictions on stonework suggest they saw a turtle in the sky. The ancient Egyptians viewed the same stars as a scarab beetle, perhaps linked to the nearby star we know as Sirius in **14. Canis Major**, which they saw as representing scarab-headed god Ra. Note that crab, turtle and scarab are all stocky creatures with a hard carapace and exoskeleton, suggesting that these different cultures each saw the same shape in this pattern of stars but imposed their own meanings.

For the Greeks, this was the crab that bit or nipped **40. Hercules** while he battled the monstrous **42. Hydra**. The great hero crushed this troublesome crab underfoot, but it was placed in the sky by Hera, a goddess with a grudge against Hercules.

INTERESTING FACTS

Ptolemy saw an entire crab in the sky, with star α Cancri on one of the two prominent front claws, and ι Cancri on the other. Some more recent depictions of the constellation show just a single claw, with ι Cancri on the arm-like limb, and α Cancri and β Cancri as opposing pincers.

'Habitable' or 'Goldilocks' Zones

One thing consistent in all life on Earth is the need for liquid water. Therefore, when we look for signs of life anywhere else but on Earth, the obvious candidates are places where liquid water may exist. It's difficult to detect that directly, but we can deduce it.

Earth orbits the Sun at a distance that is neither too hot nor too cold for water to exist. We call this distance the 'habitable' or 'Goldilocks' zone, but planets orbiting other stars do not need to be at the same distance as Earth is from the Sun to have liquid water and – potentially – life. Whether a planet may be able to sustain liquid water depends on a number of factors.

The thickness of a planetary atmosphere and what it's made of can affect the surface temperature of a planet, and thus the distance it would need to be from its star to have liquid water. Also, stars vary greatly in the amount of radiation they produce, so the habitable zone will be further out from a brighter star than it will be from one producing less radiation.

The more accurately we can calculate habitable zones around different stars, the better we can predict the chances of there being life as we know it on the planets we find there.

However, discoveries of life in places such as the Mariana Trench (the deepest part of Earth's oceans, some 11km deep) has widened our view of the possible places where life can exist. Life in the Mariana Trench gets energy not from the Sun but from thermal vents.

We're now investigating other bodies in our Solar System, such as the many moons of the outer planets, for possible signs of life.

13. CANES VENATICI, THE 'HUNTING DOGS'

Pronounced: 'KAN-iss ve-NAT-iss-eye'
Short: CVn
Brightest star: α Canum Venaticorum or Cor Caroli (RA 12h 56m, Dec. +38°19')

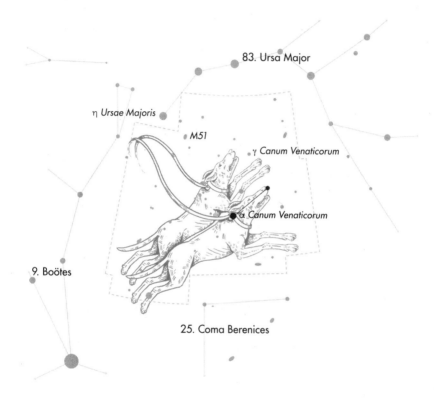

The main constellation consists of four stars. Canes Venatici is not one of the major constellations – in fact, when Claudius Ptolemy wrote his Almagest around 150 CE, he thought these faint stars were part of another constellation.

WHEN AND WHERE TO OBSERVE

This constellation is best observed during April evenings in the northern hemisphere. It is visible between latitudes +90 to −40°.

THE BRIGHTEST STARS

The brightest star, α Canum Venaticorum, is also known as Cor Caroli, or 'heart of Charles', and was named by Sir Charles Scarborough (1615–94) –

but not after himself. Scarborough was, from 1660, physician to Charles II, King of England, Scotland and Ireland, but some suggest that the star wasn't named after this Charles either, but was rather a tribute to his father, Charles I, who was executed in 1649.

When seen through a telescope, α Canum Venaticorum is a binary star (see **14. Canis Major**), the brightest component being an A-type with an apparent magnitude that varies slightly over a regular period of 5.5 days. We currently think that this happens because the star's powerful magnetic field produces very large 'starspots' – just like our Sun has patterns of relatively darker, cooler 'sunspots' on its surface.

Y Canum Venaticorum is a red giant so deeply, strikingly red – even to the naked eye – that it's also known as La Superba, the 'superb [star]'. At about 2,760K, it's also thought to be one of the coolest 'true' stars, where nuclear fusion is still continuing – but this 'cool' temperature is still hot enough to melt lead and silver!

OTHER BODIES

Known as the Whirlpool Galaxy, beautiful M51 is exactly what we tend to think a galaxy should look like. It was the first to be classified as a spiral galaxy. With binoculars, the spiral arms and bright centre are clearly visible. Through a good telescope, we can see that it is linked to – and drawing stars from – a second, smaller galaxy known as NGC 5195. M51 can be easily found by looking slightly southwest of η Ursae Majoris, which is one end of the Plough in **83. Ursa Major**.

METEOR SHOWERS

Five meteor showers are associated with this constellation, but none of them are very impressive. The most prominent, the Canum Venaticids, occur in January each year.

MYTHOLOGIES

As the stars in this constellation are not that bright, there is not a huge amount of written mythology associated with them. Yet the theme of hunting does seem to permeate from different cultures.

Ptolemy saw these stars as the *kollorobos* or wooden club carried by herdsman **9. Boötes**, while others saw them as two dogs running beside him. To the Babylonians, the stars represented stags and were incorporated into their constellation ARAKU. Early Romans seem to have seen these stars as a stag.

INTERESTING FACTS

In the 800s CE, *Almagest* was one of a number of works translated into Arabic for the Grand Library of Baghdad. For some reason, translator Hunayn ibn Ishaq (809–873 CE) changed the 'club' in this constellation to an Arabic weapon, a pole with a hook. Some say he did so because he didn't know the Greek word *kolloboros*, but Hunayn was a highly accomplished scholar and the Greek word was relatively common, so he may just have thought the Arabic pole and hook was more appropriate for describing these stars.

Gerard of Cremona (c. 1114–1187), working in Toledo in what is now Spain, translated the Arabic version of *Almagest* into Latin. He mistook Hunayn's word *kullāb* ('hook') for *kilāb* ('dogs'), and as a result of this error, later depictions of **9. Boötes** by European astronomers showed the herdsman with two dogs.

In 1687, these stars were listed as a separate constellation by the husband-and-wife team Johannes and Elisabeth Hevelius, working from their home in Danzig, Poland. They were clearly aware of the tradition that began with Gerard of Cremona's mistake, as they called the new constellation Canes Venatici, the 'Hunting Dogs'. (For more about the constellations discovered by Johannes and Elisabeth Hevelius, see **45. Lacerta**.)

14. CANIS MAJOR, THE 'GREATER DOG'

Pronounced: 'KAN-iss MAY-jer'
Short: CMa
Brightest star: α Canis Majoris or Sirius (RA 6h 45m, Dec. −16°42')

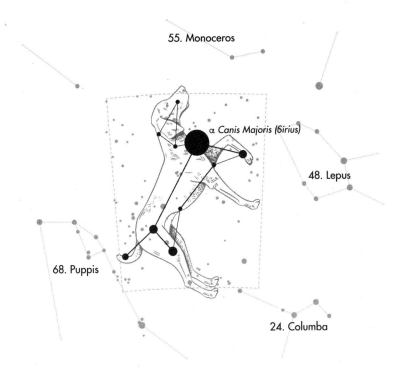

Containing the brightest star in the night sky, this constellation has captured the imagination of many different cultures around the world. The brightest star was known as the 'dog star' for centuries, and it is thought that this is where the constellation gets its name.

WHEN AND WHERE TO OBSERVE

Canis Major is prominent in winter skies in the northern hemisphere and in summer skies in the southern hemisphere. We can use the easily recognisable **60. Orion** to find bright α Canis Majoris: just follow an imaginary line westward from the three bright stars of Orion's belt.

THE BRIGHTEST STARS

α Canis Majoris is certainly distinctive: the brightest star in the sky after the Sun. In the *Iliad* by Homer, written about 750 BCE, it blazes into view in late summer, is as brilliant as polished bronze armour and is known as 'Orion's dog' because it follows on the heels of **60. Orion**, the 'Hunter'.

We now know that Sirius is a binary star some 8.7 light years from Earth, and both components are visible when seen through a telescope. Sirius A is a bright blue-white A-type star, and Sirius B is a much fainter white dwarf – the small, dense core of a former star no longer capable of nuclear fusion. Sirius B, sometimes known as the 'Pup' of the dog star, is the closest known white dwarf to Earth.

OTHER BODIES

The open cluster M41 can be found a few degrees south of Sirius. The cluster contains about 100 stars. It is thought to have a diameter of just over 25 light years and is around 200 million years old.

The emission nebula NGC 2359, also known as 'Thor's Helmet', sits some 12,000 light years from Earth and is around 30 light years in diameter. It is thought to be the result of a very hot giant Wolf–Rayet star at its centre being in the early stages of supernova.

METEOR SHOWERS

There are no significant meteor showers associated with this constellation.

MYTHOLOGIES

The Egyptians associated the star we call Sirius with their sun god Ra, and they knew that when it first appeared over the horizon, at dawn in late summer, the all-important floods would soon follow. It was often used to mark the beginning of a new year.

To the Greeks, the star became visible in the hottest part of the summer, a period still sometimes known as the 'dog days'. Some Greeks and Romans – such as Marcus Manilius, writing in about 30–40 CE – even thought (wrongly) that the bright star caused this hot period as its own energy combined with the heat from the Sun.

Ptolemy knew both the star and its constellation as the 'dog', but the star had also long been known as Sirius, from the Greek for 'scorching', in reference to its brightness. As well as different names for the star, the Greeks had different ideas about the role it played. Some thought that, as Orion's dog, it was in pursuit of either **48. Lepus**, the 'Hare', or **78. Taurus**, the 'Bull'.

Others linked the dog star to the mythical Laelaps, a hunting dog empowered by the gods to always catch its prey. In the story, Laelaps was sent out to hunt the Teumessian fox, which had also been empowered so that it could never be caught. The resulting paradox so confounded Zeus, the father of the gods, that he took mercy on Laelaps, freed him from the impossible hunt and placed him up in the sky.

In Chinese astronomy, α Canis Majoris is Tianlang, the 'celestial wolf', more dangerous than the friendly, useful hunting dog of the Greeks.

Tianlang was often associated with the threat of invasion, with other nearby stars forming a bow and arrow to hold off an attack. Many tribes across North America also associated the star with dogs or wolves – for example, to the Alaskan Inuit it is 'Moon Dog', and to different groups of Pawnee it was the 'Wolf Star' or a trickster 'Coyote'.

INTERESTING FACTS

Double or binary stars are two stars linked by mutual gravitational force and orbiting round a common centre of gravity. They really are twinned, rather than just appearing to be close together as viewed from Earth.

In some cases, the orbits of such stars are aligned with Earth in a way that means one star passes in front of the other. This results in dips in brightness at regular intervals, which we can use to calculate the orbital period, the distance of the two stars from one another and other information about these 'eclipsing binary stars'.

15. CANIS MINOR, THE 'LESSER DOG'

Pronounced: 'KAN-iss MY-ner'
Short: CMi
Brightest star: α Canis Minoris or Procyon (RA 7h 39m, Dec. +5°13')

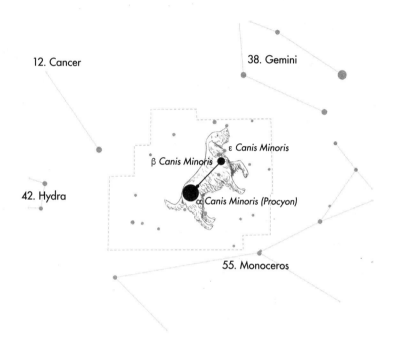

12. Cancer

38. Gemini

ε Canis Minoris

β Canis Minoris

42. Hydra

α Canis Minoris (Procyon)

55. Monoceros

*Although a relatively small constellation, Canis Minor contains the eighth brightest star seen in the night sky. Ptolemy's Almagest of around 150 CE contained the earliest known reference to this pattern of stars as a 'Lesser Dog', in comparison to the 'Greater Dog' of nearby **14. Canis Major**, both trailing the hunter **60. Orion**.*

WHEN AND WHERE TO OBSERVE
This constellation is best seen in the evenings of the northern hemisphere between January and March, and between the latitudes of +90 and –75°. To spot this constellation you can use the bright stars of **38. Gemini**, as it lies directly south of the more prominent Castor and Pollux.

THE BRIGHTEST STARS
We know now that α Canis Minoris is a binary star comprising an A-type and a smaller, fainter white dwarf some 11.3 light years from Earth. It's distinctly yellow.

OTHER BODIES

Although there are a few galaxies and nebulae within this constellation, they are generally too faint to see without professional equipment.

METEOR SHOWERS

The Canis Minorids are visible from 4 to 15 December each year, and peak on 10 and 11 December. This is not one of the major meteor showers; you might see about five meteors per hour.

MYTHOLOGIES

The Greeks knew α Canis Minoris as Procyon, meaning 'before the dog', as it is seen in the sky a few days ahead of the much brighter α Canis Majoris (the 'dog star') in **14. Canis Major**.

Other cultures had very different views of this star. To the Babylonians, α Canis Minoris was a carpenter involved in building the night sky, while to the Chinese, along with ε Canis Minoris and β Canis Minoris, it made up the 'South River'.

In Alaska, Canada and Greenland, α Canis Minoris can appear low in the sky and deep blood-red in colour, due to the effect of Earth's atmosphere. As a result, Inuit people have associated it with the story of a thief, Sikuliarsiujuittuq, who was lured out onto the frozen sea and murdered.

The Greeks also linked the star to violent death. To some, it was Maera, the dog owned by murdered winemaker Icarius and his daughter (see **9. Boötes**). In his book *Skywatching*, David Levy says the Greeks also saw this star as one of Actaeon's hounds. According to legend, Actaeon was out hunting with his dogs when he spotted a beautiful woman bathing naked in a pond. Unfortunately for Actaeon, the woman caught him watching and, even more unfortunately for him, she was the goddess Artemis. As punishment, she transformed him into a stag, and he was then hunted and killed by his own dogs.

INTERESTING FACTS

The star α Canis Minoris also has positive associations. It appears under the motto *Ordem e Progresso* ('order and progress') on the national flag of Brazil. Brazil is a republic comprising 26 states and one federal district, represented by the 27 stars on the flag, which show the night sky above Brazilian city Rio de Janeiro on 15 November 1889, the day that Brazil's first republic was established. It's an example of modern culture linking itself and its aspirations to the stars, just as the ancients did.

The national flag of Brazil depicts the brightest 27 stars from nine constellations:

- 5 from **14. Canis Major**
- 1 from **15. Canis Minor**
- 1 from **17. Carina**

- 5 from **30. Crux**
- 2 from **42. Hydra**
- The 'South Star' in **58. Octans**
- 8 from **73. Scorpius**
- 3 from **81. Triangulum Australe**
- 1 from in **86. Virgo**

The arcing text of the motto follows the path of the ecliptic (see 'Constellations of the Zodiac' in Chapter 2).

16. CAPRICORNUS, THE 'SEA GOAT'

Pronounced: 'CAP-rih-CORN-us'
Short: Cap
Brightest star: δ Capricorni or Deneb Algedi (RA 21h 47m, Dec. −16°7')

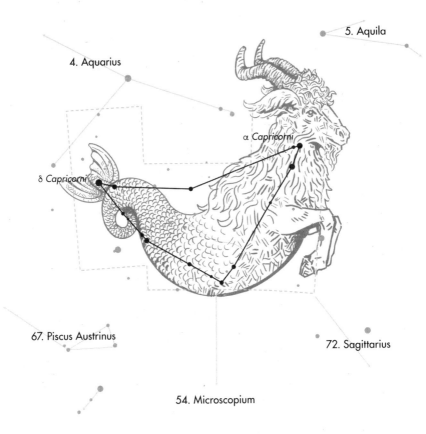

The name of this constellation is familiar because it is in the zodiac. This is a rather unlikely hybrid creature, half-goat and half-fish. Capricornus is the smallest constellation in the zodiac but about average in size when compared with the other constellations.

WHEN AND WHERE TO OBSERVE
Capricornus appears in the night sky from June through to October in the southern hemisphere and from August until November in the northern hemisphere. In the northern hemisphere, it is best seen on mid-September evenings.

THE BRIGHTEST STARS

α Capricorni is a double star: the two component stars are not actually close together, but they look that way as seen from Earth. It's just possible to see both components with the naked eye. The Arabic name for this star, Algedi, means 'billy goat' and was sometimes used to mean the whole constellation.

The slightly brighter δ Capricorni – or Deneb Algedi, the 'tail of the billy goat' – is a binary star comprising a white A-type twice the mass and almost twice the diameter of the Sun, and what's either a G-type or K-type star a little smaller than the Sun. Two further stars seem to be part of the same system, but only as viewed from Earth – so this is a double-double star with one pair a binary!

OTHER BODIES

The spiral galaxy NGC 6907 can be seen in this constellation. The Messier object M30 (NGC 7099) is a globular cluster some 30,000 light years from Earth, but its chains of stars can be observed with a relatively small telescope.

METEOR SHOWERS

A number of meteor showers are associated with this constellation. The most prominent is the Alpha Capricornids, which are the result of comet 169P/NEAT, which fell apart around 4,000 years ago. The shower can be seen between 15 July and 10 August. Although only producing around five meteors per hour now, in the distant future this number is likely to go up as Earth's path drifts into the denser areas of the debris field.

MYTHOLOGIES

To the ancient Sumerians this star combination was known as SUHUR-MASH-HA – the 'goat-fish' – and Babylonian sources from around 1000 BCE refer to it as the 'goat-fish', too.

To them, the Sun was positioned in front of this constellation on the shortest day of the year – the winter solstice, around 21 December. We still refer to the Tropic of Capricorn: a line of latitude circling Earth that is the most southerly point at which the Sun is directly overhead at midday, which happens on this same December solstice and currently occurs 23°26' south of the equator (see 'Tropic of Cancer', page 73). However, slight movement in the axis of Earth's rotation over the centuries means that today, when this solstice takes place, the Sun is in front of **72. Sagittarius**.

The ancient Greeks called this pattern of stars Aigokeros, 'goat' – perhaps without the fish parts. In one Greek myth, these stars were Amalthea, the goat that nursed Zeus when he was an infant, hiding him from his father Kronos, head of the Titans.

They also associated this goat-shaped pattern of stars with Pan, the god of the countryside. Pan was himself another unlikely hybrid, with the head and body of a man and the legs and hooves of a goat. One legend that could possibly explain the strange combination of goat and fish involves Pan escaping the monstrous Typhon by hiding himself in the river Nile and transforming the submerged part of his body into a fish. But this half-goat, half-fish story surely owes something to the earlier, Babylonian sense of this pattern of stars. (For a similar story, see **66. Pisces**.)

The earliest surviving reference to this constellation being called Capricornus – from the Latin for 'horned goat' or 'horns like those of a goat' – is in Ptolemy's *Almagest*, written about 150 CE.

INTERESTING FACTS

On 23 September 1846, German astronomer Johann Galle (1812–1910) trained his telescope on δ Capricorni and, just as he'd hoped, saw a blue-white object. Other astronomers had seen this 'star', but not in the same place in the night sky. Galle was the first person to see it knowing its true nature: that it is a planet orbiting our Sun. Today, we know it as Neptune.

Tropic of Cancer

This imaginary line of latitude around Earth is 23°26' north of and parallel to the equator. It is the most northerly point on Earth at which the Sun is directly overhead at midday, which occurs on the longest day in the northern hemisphere – the 'summer solstice' – around 21 June each year. This and the corresponding Tropic of Capricorn, 23°26' south of and parallel to the equator, are often clearly marked on maps of Earth, so they are key to navigation. The relatively warm area of Earth between the two lines is known as the tropics.

17. CARINA, THE 'KEEL'

Pronounced: 'kuh-REE-nuh'
Short: Car
Brightest star: α Carinae or Canopus (RA 6h 23m, Dec. −52°41')

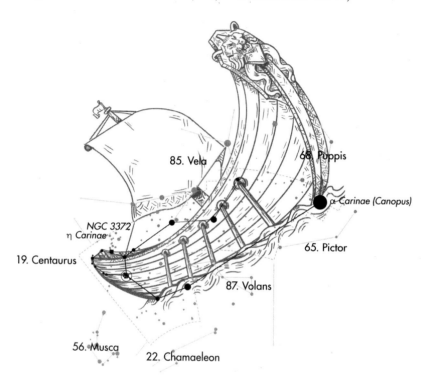

Even though it contains one of the brightest stars visible in the sky, Carina is more famous for the nebulae that can be seen within its boundaries. This constellation is also better understood in the context of several of its neighbours.

*There used to be a single, huge constellation depicting the Argo (see 'Mythologies', below), but its sheer size made it impractical to work with, so it was divided up by the French astronomer Nicolas-Louis de Lacaille in 1763. He suggested three smaller constellations representing the components of the ship. Carina is the keel, the strong, hard 'backbone' of the structure that ran front to back along the bottom of the hull, to which the ribs of the boat were attached. The constellation **68. Puppis** is the ship's poop deck and **85. Vela** its sails. Lacaille then added a fourth component: **69. Pyxis** is the ship's compass. (For more about Lacaille and the other constellations he created, see **2. Antlia**.)*

WHEN AND WHERE TO OBSERVE
Carina can be seen in the southern hemisphere from latitudes south of +20°, where it is best observed on March evenings.

THE BRIGHTEST STARS
α Carinae is the second brightest star in the night sky, after the Sun and α Canis Majoris, or Sirius, in **14. Canis Major**. It's a white A-type giant, eight times more massive than the Sun, 71 times its diameter and 10,000 as bright.

This star is known as Canopus, which is odd. Canopus was the pilot of a great ship in Greek mythology – but not of the *Argo*. So who was he and how did he get linked to this star?

In Homer's *Iliad*, written around 750 BCE and thought to be based on real events that occurred in around 1200 BCE, Canopus was pilot to King Menelaus of Sparta, whose wife was Helen of Troy. When Helen left Menelaus, the king and his brother led a fleet of 1,000 ships to Troy to recover her, beginning the ten-year Trojan War.

Canopus died and was buried near the coast of Egypt, where Menelaus left a memorial to his friend. The Greeks knew one ancient Egyptian port as Canopus, presumably thinking it was where the legendary figure had died, but some think they misheard the local name Kahi Nub, Coptic for 'golden world' – a description of the golden colour α Carinae has when it's near the horizon, an effect of Earth's atmosphere.

OTHER BODIES
The large, diffuse Carina Nebula (NGC 3372) was discovered by Lacaille around the same time that he reconfigured the Argonauts' unwieldy ship. This nebula has a diameter of more than 200 light years and sits around 8,500 light years away from Earth. Images of this body show beautiful coloured wisps of dust and gas, a maelstrom of star birth and death.

METEOR SHOWERS
The most prominent meteor shower in this constellation is the Eta Carinids, which peak around 21 January, with about two or three meteors per hour.

MYTHOLOGIES
To the ancient Greeks, the huge constellation of which Carina is a part was Argo Navis, the ship *Argo* sailed by the legendary Jason who, with his crew of Argonauts, went on a quest to find the Golden Fleece. The Greek poet Aratus (c. 315–240 BCE) noted in his *Phenomena* that the *Argo* moved through the night sky as if sailing backwards.

Greek historian Plutarch (c. 46–119 CE) thought that Argo Navis had been inherited from the Egyptians who, he said, had seen the same pattern of stars as the Boat of Osiris, their god of the dead. Yet no evidence of this has been found in the archaeology of Egypt.

Some have suggested that the Greeks actually inherited this boat-shaped pattern of stars from the Babylonians, whose *Epic of Gilgamesh* (written c. 1600–1200 BCE) features a man instructed by the gods to construct an enormous ship, with which he then saved his people and their livestock from a devastating flood. But again, no evidence has been found to show that the Babylonians linked that famous story to this particular pattern of stars. Argo Navis may have been an entirely Greek creation.

Most of the stars that make up this constellation are barely visible in China. However, the bright star Canopus was included in their constellation Vermilion Bird of the South, representing the element of fire.

For the Polynesians, the star Canopus was again the main attraction; they had many names for this star, including 'chief of the southern sky', 'high born' and 'stand alone'.

INTERESTING FACTS

η Carinae is a binary star system that may contain more than two stars and is one of the most massive stars in our galaxy. It was first catalogued by English astronomer Edmond Halley (1656–1742). Then, in the early decades of the nineteenth century the star became brighter and brighter, peaking between 11 and 14 March 1843, when it was observed to be second only in brightness to Sirius (in **14. Canis Major**). Since then, it has dimmed and was impossible to see with the naked eye during the twentieth century. Since 1940, it has started to grow in brightness again, becoming visible to the naked eye once more.

One star in the η Carinae system seems to be going through its death throes, increasing in brightness and then settling back down. This behaviour indicates that the star is likely to go supernova in the near future. As this star is 7,500 light years away from us, we will only mildly feel the effects of such a huge explosion of energy. Protected by Earth's atmosphere, we're unlikely to feel any ill effects at all on the surface, but objects in orbit may be more vulnerable.

18. CASSIOPEIA, THE 'SEATED QUEEN'

Pronounced: 'CASS-ee-oh-PEE-uh'
Short: Cas
Brightest star: α Cassiopeiae or Schedar (RA 0h 40m, Dec. +56°32')

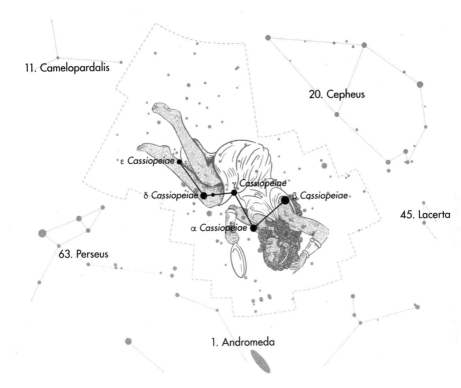

Cassiopeia is one of the larger, more easily spotted constellations due to its distinctly W-shaped asterism. In Almagest, *written about 150 CE, Greek-Egyptian astronomer Claudius Ptolemy identified this constellation as a seated figure: Cassiopeia, an Ethiopian queen from mythology and the wife of* **20. Cepheus.**

WHEN AND WHERE TO OBSERVE

This constellation can be seen all year round in the northern hemisphere, roughly above latitude +34°; south of this, it appears seasonally. It is located on the far side of α Ursae Minoris, also known as Polaris or the North Star (in **84. Ursa Minor**), which can be identified from the easy-to-find Plough (in **83. Ursa Major**).

THE BRIGHTEST STARS

α Cassiopeiae is also known as Schedar, from the Arabic for 'breast', a K-type red giant roughly four times the mass of the Sun, 45 times the diameter and 800 times as bright. Its apparent magnitude is 2.24.

β Cassiopeiae, or Caph (the 'hand'), is an F-type yellow-white star whose brightness varies between apparent magnitudes of 2.25 and 2.32 every 2.5 hours, the result of its size and the fact it has nearly burnt through its supply of hydrogen.

γ Cassiopeiae is a more noticeably variable star, ranging between apparent magnitudes of 1.6 and 3.0 in no fixed pattern. This means it is sometimes brighter than both α Cassiopeiae and β Cassiopeiae. The variation in brightness is caused by the star rotating very fast and flinging off material from its surface!

δ Cassiopeiae, or Ruchbah (the 'knee'), has an apparent magnitude of 2.68, but that fluctuates very slightly, by less than 0.1 in a regular cycle of 759 days. This is because it is a binary star, and one component sometimes eclipses the other as seen from Earth, resulting in less light being seen. Its brighter component is an A-type star that has exhausted its hydrogen and is in the process of growing into a giant.

ε Cassiopeiae, or Segin (where this name comes from is unclear), is a blue-white B-type with a surface temperature of 15,174K. It's surrounded by a 'shell' of gas that the star has thrown off because it rotates so quickly – though with only a very minor impact on its brightness.

OTHER BODIES

This constellation sits on a background rich in galactic objects. Two Messier objects are found in this constellation: M52 and M103 are open clusters sitting 5,000 and 10,000 light years from Earth, respectively.

Sometimes known as Caroline's Rose, NGC 7789 is another open star cluster, some 7,600 light years distant from us. The name comes from British astronomer Caroline Herschel, who discovered it in 1783.

Sitting near M52, NGC 7635 is an emission nebula commonly known as the Bubble Nebula. Visible with a telescope of eight inches or more, its beautiful wispy colours were captured in 2016 by the Hubble Space Telescope as part of its 26th birthday celebrations.

Cassiopeia A is the strongest radio source observed outside our Solar System. This supernovae remnant is ten light years across.

METEOR SHOWERS

φ-Cassiopeid is a minor meteor shower, generally active between 1 and 8 December, peaking on 6 December.

MYTHOLOGIES

According to legend, Cassiopeia angered the gods by boasting about the beauty of her daughter, **1. Andromeda** (see that entry for the story, which also involves **20. Cepheus, 21. Cetus, 62. Pegasus** and **63. Perseus**). In later versions of the legend, the gods had Cassiopeia tied or chained to her throne and sent wheeling round the North Star as punishment. As she wheeled round, she would – when seen from Greece – partly drop behind the horizon over the sea, as if being 'dunked' in the water.

Cassiopeia may have been a real person and the story based on some real event, but no archaeological evidence has yet been found for her. The ancient Greeks couldn't even agree where exactly Cassiopeia reigned. Although today there's a country called Ethiopia in East Africa, the word means 'dark-skinned' and the ancient Greeks often used it to mean Africa more generally. Some sources included the southeastern coast of the Mediterranean and what is now Israel and Jordan as part of this ancient Ethiopia.

The stars in this constellation are found in three of the four areas representing mythical creatures found in in Chinese astronomy: the Purple Forbidden Enclosure, the Black Tortoise of the North and the White Tiger of the West.

An Arabic constellation called the Camel also incorporates some of the stars associated with Cassiopeia. The full Camel overlaps with stars in **63. Perseus** and **1. Andromeda**.

INTERESTING FACTS

γ Cassiopeiae was used by early American astronauts as a navigational reference point. Virgil 'Gus' Grissom (1926–67) even nicknamed it 'Navi', which is still sometimes used.

In the film *Serendipity* (2001), a character played by John Cusack tries to impress a woman by telling her about the constellation Cassiopeia. Serendipitously, in the film *The Sure Thing* (1985), a character played by John Cusack also tries to impress a woman by telling her about the same constellation – who knew constellations could make such good flirting lines!

19. CENTAURUS, THE 'CENTAUR'

Pronounced: 'SEN-tor-us'
Short: Cen
Brightest star: α Centauri or Rigel Kentaurus (RA 14h 39m, Dec. −60°50')

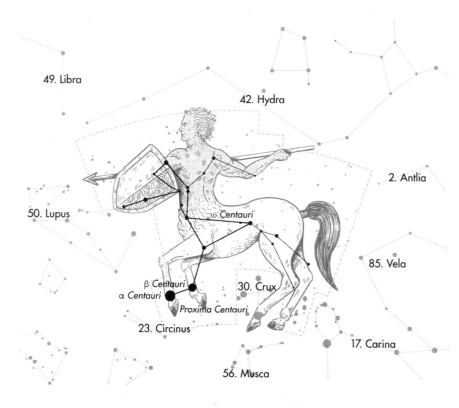

This constellation is made up of 17 stars. The centaur is a creature from ancient Greek myth. It's often said that the myth arose when ancient people first saw someone on horseback and assumed it was all one being, half-human and half-horse. However, there's no historical evidence to back this theory, and many mythological creatures are animal hybrids – for example, **62. Pegasus**, a horse with wings and the ability to fly like a bird, or **16. Capricornus**, the rather unlikely 'sea goat'.

This constellation can be used to find the Southern Cross. The two stars at the base of this constellation serve as pointers to the Southern Cross, which lies under the centaur's rear legs.

WHEN AND WHERE TO OBSERVE

This southerly constellation is visible at latitudes between +25 and −90°. It is observable in the northern hemisphere, and is best observed in the southern hemisphere during May.

THE BRIGHTEST STARS

Sometimes known as Rigel Kentaurus, the 'foot of the centaur', α Centauri is the third brightest star in the night sky (after Sirius in **14. Canis Major** and Canopus in **17. Carina**). It is really a triple system, with three stars in orbit round one another. The two brightest are both yellow, one G-type and one K-type, and rather like our Sun. The third component, Proxima Centauri, can only be seen with a telescope. It's a red dwarf just 215,000km in diameter – about 14% that of the Sun. At 4.25 light years away, it is also the closest star to the Sun. Proxima b, the closest exoplanet to Earth, orbits this star.

β Centauri is another bright star, easily visible with the naked eye; it is also a triple system. What's more, α Centauri and β Centauri are known as the Southern Pointers and can be used to find the south celestial pole, which is of great use to navigators (see **30. Crux**).

OTHER BODIES

Centaurus contains five galaxies, five nebulae, eight star clusters and one supermassive black hole.

ω Centauri is an oval-shaped globular cluster of perhaps as many as 10 million stars. At 17,000 light years from Earth, it's not quite the closest such cluster, but it is more readily visible to the naked eye than NGC 6397 in **6. Ara** and M4 in **73. Scorpius**.

Centaurus A is a giant elliptical galaxy sitting some 12 million light years away. At the centre of the galaxy is a supermassive black hole with a mass of 55 million Suns.

METEOR SHOWERS

There are at least three meteor showers associated with this constellation. The Alpha Centaurids are the most prominent but still a fairly weak shower, which peaks at about three meteors an hour in early February each year.

MYTHOLOGIES

The Babylonians saw in this pattern of stars a half-human, half-bison creature that they knew as MUL.GUD.ALIM. We're not sure exactly when or why this became a centaur, but astronomer Eudoxus of Knidos is thought to have referred to this constellation as such in now lost writings from about 400–350 BCE (see **5. Aquila**), as did Aratus a century later. Centaurus is one of the 48 constellations in Ptolemy's *Almagest*, written about 150 CE.

In some accounts, the constellation is a particular centaur. Whereas other centaurs were thought to be wild and dangerous, Chiron was wise and nurturing. His students included legendary hero **40. Hercules**, who accidentally wounded him. Since Chiron was immortal, he was left in endless pain – until the gods took pity on him and placed him in the night sky. Some ancient Greek sources also claimed that **6. Ara** is Chiron's altar.

INTERESTING FACTS

Given its proximity to us (hence the name), it's especially exciting that we now know of at least two planets in orbit around Proxima Centauri, one of which is within the star's habitable zone (see page 61) – though the fact that it is a red dwarf may mean it does not radiate enough energy to support life.

20. CEPHEUS, THE 'KING'

Pronounced: 'SEE-fee-us'
Short: Cep
Brightest star: α Cephei or Alderamin (RA 21h 18m, Dec. +62°35')

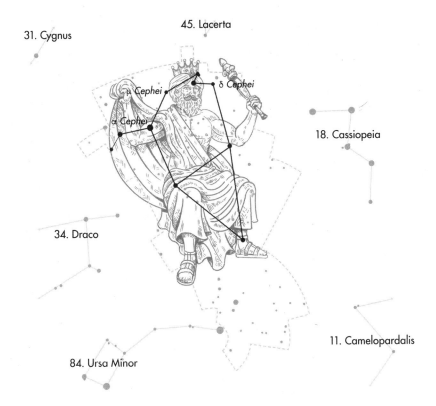

This legendary king of Ethiopia was the father of **1. Andromeda** (see that entry for their story). The constellation is one of the 48 listed by Greek-Egyptian astronomer Claudius Ptolemy in Almagest, written about 150 CE. It can be found by star hopping from the distinctive 'W' asterism in **18. Cassiopeia**; the centre of the 'W' can be used as an arrow that points to Cepheus.

WHEN AND WHERE TO OBSERVE
Given its proximity to the north celestial pole, Cepheus is visible throughout the year in the northern hemisphere but is most prominent on November evenings. It is viewable from latitudes between +90 and −10°.

THE BRIGHTEST STARS

The white, A-class α Cephei is also known as Alderamin, a shortened version of an Arabic phrase meaning the 'right arm [of Cepheus]'. It varies slightly in brightness, but not nearly so noticeably as δ Cephei, a group of four stars in close proximity to each other that varies in apparent magnitude from 3.48 to 4.37 every 5.37 days (see 'Interesting facts', below, for why this is useful to us).

μ Cephei is a red supergiant visible to the human eye. It's often known as Herschel's Garnet Star, after the famous astronomer William Herschel (1738–1822), who admired its 'fine' red colour. We now know that this colour is a sign that the star is nearing the end of its life. It has already exhausted its supplies of hydrogen and is now fusing helium instead. When it uses up its fuel, the star's core will collapse and produce a supernova. It's thought μ Cephei is massive enough that this might create a black hole (see page 13).

OTHER BODIES

Cepheus is not known to contain any Messier objects, but it does have some deep-space objects, such as NGC 6946, known as the Fireworks Galaxy. This can only be seen with larger telescopes and has been observed to contain more than nine supernovae – this is more than in any other galaxy.

METEOR SHOWERS

Although there are five meteor showers associated with this constellation, none of them are very prominent.

MYTHOLOGIES

Ptolemy described Cepheus as wearing a tiara. It's been suggested that he meant the kind of headdress worn by kings of Persia, and that this is evidence that Cepheus' kingdom was really to the east of ancient Greece, and not further south in Africa where the modern country of Ethiopia is located (see also the entry on Cepheus' wife, **18. Cassiopeia**).

INTERESTING FACTS

The change in the apparent magnitude of δ Cephei is so regular and reliable that the star has given its name to the Cepheid variables, a type of star that rapidly pulsates in a stable, regular way. The relationship between the apparent brightness and the pulsation period allows us to calculate the actual brightness of the star (not just how it appears to us from Earth), and thus its distance. With this standard to guide us, we can then calculate the distances of other objects in the sky.

21. CETUS, THE 'SEA MONSTER'

Pronounced: 'SEE-tus'
Short: Cet
Brightest star: β Ceti or Diphda (RA 0h 43m, Dec. −17°59')

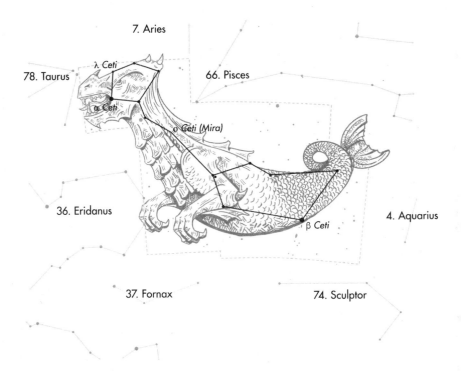

7. Aries

λ Ceti

78. Taurus

66. Pisces

α Ceti

ο Ceti (Mira)

36. Eridanus

4. Aquarius

β Ceti

37. Fornax

74. Sculptor

Cetus is the fourth largest constellation in the sky. It is thought that the Babylonians saw this pattern of stars as a whale. That tradition was continued as Cetus is the Latin word for 'whale'. The constellation can be found by following the three stars in the belt of **60. Orion** *up to the distinctive 'V' shape of* **78. Taurus***; this 'V' shape then points towards Cetus.*

WHEN AND WHERE TO OBSERVE

Visible from latitudes south of +70°, it's most visible in the northern evening sky during November. This constellation lies close to the ecliptic, so it can be obscured by the Moon, depending on its phase. The best months to spot the constellation in the southern hemisphere are January and February.

THE BRIGHTEST STARS

β Ceti is a bright orange giant somewhere on the border between being a G-type and a K-type star. The name Diphda comes from an Arabic phrase meaning 'second frog' (the first frog being Fomalhaut in **67. Piscis Austrinus**). It also used to be known as Deneb Kaitos, from an Arabic phrase meaning 'tail of Cetus'.

Red giant α Ceti is slightly less bright. It's commonly known as Menkar, the 'nostrils', though Ptolemy described this star as being on the creature's jaw. Confusingly, λ Ceti is also known as Menkar and is in the right position for the creature's nostrils.

OTHER BODIES

The Messier object M77 is found in this constellation. It sits 47 million light years from us, and it is a good galaxy to observe as it is face on to us, so the detail of its spiral arms can be enjoyed. The rather haunting Skull Nebula (NGC 246) can also be found in this constellation.

In August 2022, the Hubble Space Telescope discovered in Cetus a star initially classified as WHL0137-LS and since known as Earendel, from an Old English word for 'morning star'. At some 28 billion light years away, this is the most distant and earliest star yet detected – and almost twice the distance from us of the previous holder of the record, the star Icarus in **46. Leo**. Earendel is so distant that we can only see it because of gravitational lensing: by chance, the mass of a huge galaxy cluster between the star and us magnifies what we can see by 2,000 times!

METEOR SHOWERS

Three annual meteor showers are associated with this constellation. The Eta Cetids and Omicron Cetids are best seen between 7 May and 9 June each year.

MYTHOLOGIES

To the ancient Greeks, Cetus was the sea monster slain by **63. Perseus** to save the life of **1. Andromeda** (see her entry for the story). This monster was often depicted as having the head and front legs of a land animal (which animal could vary), a fish-like body and a coiled tail.

Cetus is associated with other nearby constellations: Andromeda's parents, **18. Cassiopeia** and **20. Cepheus**, who are part of the same legend, as is Perseus' winged horse, **62. Pegasus**. Cetus is also in the middle of a 'sea' of water-based constellations: **4. Aquarius**, **36. Eridanus**, **66. Pisces** and **67. Piscis Austrinus**.

INTERESTING FACTS

The best-known star in the constellation is o Ceti, which is sometimes but not always visible to the naked eye. We think this star was observed by astronomers in ancient Babylon, China and Greece, but we're not sure whether they noticed its unusual properties.

When German pastor David Fabricius (1564–1617) observed this star on several occasions in 1596, it was less bright each time. Initially, he thought it must be another example of a nova – what we now know to be a dying star – but then in 1609 Fabricius spotted o Ceti again and it had grown brighter. He was probably the first person ever to notice such a phenomenon.

By 1638, o Ceti was recognised as the first known 'variable' star, its brightness rising and falling in a more or less regular cycle, in this case lasting about 300 days. This variability was thought to be so astonishing that in 1662 Polish astronomer Johannes Hevelius named the star Mira – Latin for 'wonderful' – and the name is still in use today. It is a binary star system comprised of a pulsating red giant and a fainter white dwarf.

22. CHAMAELEON, THE 'CHAMELEON'

Pronounced: 'kah-MEAL-ee-on'
Short: Cha
Brightest star: α Chamaeleontis (RA 8h 18m, Dec. −76°55')

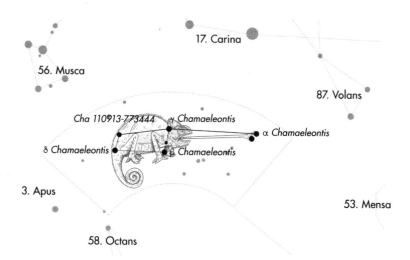

As the 79th largest constellation out of the 88, Chamaeleon is small and
has no bright stars within it. It is named after the lizard found in Africa,
Madagascar and across South Asia. The name is attributed to the Dutch
astronomer Petrus Plancius, even though he never viewed the constellation
himself or likely even saw a live specimen of the creature he named it after.

WHEN AND WHERE TO OBSERVE
This constellation sits close to the south celestial pole. Chamaeleon is best
seen in southern skies during April.

THE BRIGHTEST STARS
α Chamaeleontis is a white F-type star and β Chamaeleontis a blue-white
B-type star. By contrast, γ Chamaeleontis is a K-type red giant – the colour is
distinct compared to its two neighbours. The other bright star of the constellation is δ Chamaeleontis, a double star comprising two components that
look close together as seen from Earth but are not actually near one another.

OTHER BODIES
Visible only with high-end, professional telescopes, this constellation is
home to a very strange object. Cha 110913–773444 is either a very small
type of star called a brown dwarf, or a 'rogue planet' – a planet out on its

own in space rather than orbiting a star. There also appears to be a disc of material around this object: if it's a brown dwarf, it has planets; if it's a planet, it has moons. Whichever is the case, it's an extremely unusual object – so far, we've found nothing like it!

METEOR SHOWERS
There are no known meteor showers in this constellation.

MYTHOLOGIES
Due to the size of this constellation and the lack of bright stars contained within it, there are no known mythologies around these stars. However, in one depiction of Chamaeleon, thought to have been created by Dutch cartographer Jodocus Hondius (1563–1612), the lizard is shown sticking its tongue out, trying to catch the fly represented by neighbouring constellation **56. Musca**.

INTERESTING FACTS
In September 1595, a fleet of four Dutch ships arrived at Madagascar in the Indian Ocean. It had been a rough voyage, and 71 of the 248 crew members had died of scurvy (a nasty disease we now understand to be caused by a lack of vitamin C in the diet). Exhausted, the crew stayed on the island for four or five months to recover, repair their ships and gather essential supplies.

Three sailors – Pieter Dirkszoon Keyser, Frederick de Houtman and Vechter Willemsz – also carefully recorded the stars. Keyser had been trained in mathematics and astronomy by Dutch mapmaker Petrus Plancius, who was keen to map the then-uncharted southern night skies. The hope was that accurately mapping these stars would give the Dutch sailors a navigational advantage over their rivals and make long voyages easier.

The fleet set sail again in February 1596, and after four months reached Sumatra in western Indonesia. But conditions remained tough, and on Sumatra the crew ran out of drinking water and other provisions. Keyser died that September.

A year later, de Houtman and the surviving members of the crew arrived home in the Netherlands, and the observations he had made with Keyser and Willemsz were delivered to Petrus Plancius. From this data, Plancius created 15 new constellations, initially depicted on a celestial globe of 1598 (or possibly the year before), and then in 1603 included in the celestial atlas published by Johann Bayer (of the Bayer designation, see page 21). (For more about the constellations created by Plancius, see **30. Crux**.)

Chameleons are famously able to change the colour and pattern of their skin. This was originally thought to be used in camouflage for hiding the creature from predators, so perhaps that's what inspired the name of this small, faint and hard-to-see constellation. We now know this colour changing is also used to communicate with other chameleons and to respond to changes in temperature.

23. CIRCINUS, THE 'COMPASS'

Pronounced: 'SER-sin-us'
Short: Cir
Brightest star: α Circini (RA 14h 42m, Dec. −64°58')

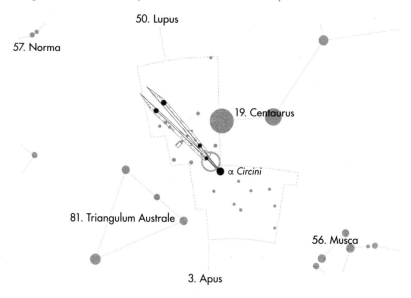

*Being the fourth smallest constellation and having no bright stars, this is another of the fill-in constellations created by French astronomer Nicolas-Louis de Lacaille (see **2. Antlia**). Circinus is Latin for 'compass' – comprising two legs connected by a hinge, it's an instrument used in technical drawing to produce circles and arcs. Such compasses have a wide range of uses in mathematics and art.*

WHEN AND WHERE TO OBSERVE
Circinus is a relatively small, faint constellation visible in southern skies, and it is best seen on July evenings.

THE BRIGHTEST STARS
α Circini forms the 'hinge' of the compass and is a white A-type star that we know to be rapidly oscillating. It's also a binary star, but its orange, K-type companion can only be seen with a telescope.

OTHER BODIES
Although there are no Messier objects associated with this constellation, its position on the backbone of our Milky Way galaxy means there are a number of objects of interest in this constellation.

The Circinus Galaxy (ESO 97-G13) has an extremely bright nucleus and probably a supermassive black hole at its centre. It is 260 light years across and sits some 13 million light years from our galaxy. It is one of the closest active galaxies to our own.

The constellation Circinus also contains a number of open clusters.

METEOR SHOWERS

Surprisingly for such a small constellation, there is a meteor shower associated with Circinus. The Alpha Circinids peak around 4 June each year, when a respectable 15 meteors per hour can be seen.

MYTHOLOGIES

As with most of the constellations created by Lacaille, there are no known mythologies about these stars.

INTERESTING FACTS

French astronomer Nicolas-Louis de Lacaille, who created this constellation in 1756, used a compass like this in his work as a surveyor. It's one of 14 constellations he created, most of them representing technical equipment of a similar kind (see **2. Antlia**), including another sort of compass: **69. Pyxis**, the 'Mariner's Compass', which points due north and is used by sailors to navigate.

24. COLUMBA, THE 'DOVE'

Pronounced: 'koh-LUM-buh'
Short: Col
Brightest star: α Columbae or Phact (RA 5h 39m, Dec. −34°4')

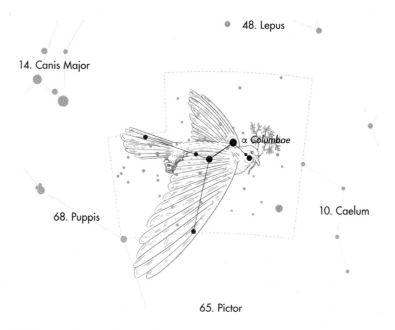

This is another small constellation introduced by Dutch astronomer Petrus Plancius in the late sixteenth century (see **30. Crux**), who originally named it Columba Noachi, meaning 'Noah's dove'. This has since been simplified.

WHEN AND WHERE TO OBSERVE
Columba sits between the two brightest stars in the night sky, Sirius (in neighbouring **14. Canis Major**) and Canopus (in nearby **17. Carina**); trace a line between them to find this constellation. The fact that these neighbours are so dazzling may explain why it took so long for a pattern to be applied to these relatively faint stars.

THE BRIGHTEST STARS
α Columbae is a blue-white B-type some 261 light years from Earth.

OTHER BODIES
Blue-coloured μ Columbae is one of very few O-types that are visible to the naked eye, and it is approximately 1,300 light years away from us. It doesn't look like it's moving, but it is – and fast. Studies of radial velocity

(see 'Exoplanets', page 16) reveal that this and another star, AE Aurigae in **8. Auriga**, are moving away from each other at more than 200km/s. To put that in context, that's like travelling from Birmingham to London in one second! What can make them move at such a rate? Tracing their trajectory back to a single point, we think these 'runaways' were part of a collision of binary stars some 2.5 million years ago and they've been hurtling off in either direction ever since.

METEOR SHOWERS
There are no known meteor showers in this constellation.

MYTHOLOGIES
To Chinese astronomers, the six brightest stars in this constellation depicted – in pairs – a farmer, his son and his grandson. According to Ian Ridpath's book *Star Tales*, another star in the north of the constellation (we're not sure which) was identified as a poo, from the 'celestial toilet' made up of the four brightest stars in **48. Lepus**!

As we said in the introduction, Petrus Plancius named this constellation Columba Noachi, Latin for 'Noah's dove'. According to the Biblical story, Noah built a huge boat – the Ark – so that he, his family and a lot of animals could survive a flood that engulfed the whole world. After 40 days and nights of rain, Noah and his crew had to spend another 150 days cooped up on the boat with no sign of land. Imagine that mixture of lockdown and seasickness, and the smell of all those animals!

No wonder, then, that when a dove flew off and returned with a branch from an olive tree, everyone on the Ark breathed a great sigh of relief. The dove just wanted to build itself a nest, but to Noah and his family the branch was a sign that the waters were at last receding. Soon enough, the Ark settled on dry land.

INTERESTING FACTS
Noah and his Ark is a classic story, one I'm sure you've heard before. So why did Plancius apply it to these particular stars?

Well, we're not really sure. It doesn't fit with any of the other constellations and there's no pattern of stars representing Noah. In 1613, Plancius claimed that a nearby constellation, Argo Navis – a boat from Greek mythology – was really Noah's Ark and thus linked to the dove Columba, but that idea never really caught on.

The huge Argo Navis was later broken up into smaller constellations (see **17. Carina**). Ironically, some more recent histories have suggested that Columba would fit with the Greek myth of the *Argo*, as the dove released by the heroic Jason.

But perhaps Plancius was inspired to name this constellation after the Arabic name for the constellation's brightest star: Phact means 'ring-dove', a domesticated species of dove.

25. COMA BERENICES, 'BERNICE'S HAIR'

Pronounced: 'KOE-muh BEH-ruh-NICE-eez'
Short: Com
Brightest star: β Comae Berenices (RA 13h 11m, Dec. +27°52')

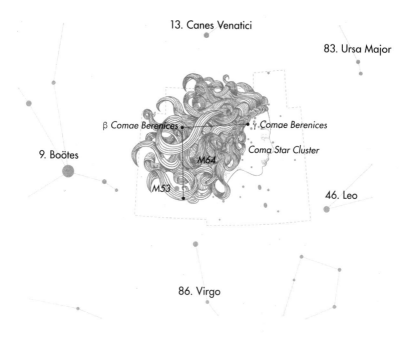

This official constellation is relatively new, but the lore behind it is extremely old. These stars were formally part of the tuft on the end of the tail of **46. Leo**. But in the sixteenth century the tuft was promoted to a full constellation: still hair, but now belonging to Berenice II (c. 267–221 BCE). Berenice was queen of Cyrenaica – in what is now Libya – from about 250 BCE, and also queen of Egypt from 246 BCE. This is the only constellation named after someone who we have evidence really existed.

WHEN AND WHERE TO OBSERVE

This relatively faint constellation can be seen from both the northern and southern hemispheres, north of −70° latitude. It's most prominent during May.

THE BRIGHTEST STARS

The brightest star, β Comae Berenices, has an apparent magnitude of only 4.26, so it is often difficult to see at all with the naked eye in light-polluted skies. It's a yellow G-type star, slightly larger, brighter and younger than our Sun.

OTHER BODIES

The tresses of Berenice's hair are really an open cluster of some 40 bright stars in a distinctive 'V' shape, though the brightest of these – orange K-type γ Comae Berenices – has an apparent magnitude of only 4.36.

Binoculars or a good telescope reveal many striking wonders in this part of the sky, including a good example of a globular cluster (M53) and the Black Eye Galaxy (M64), so named because the light from its stars partly is obscured by an enormous, dark cloud of dust in such a way that it resembles an eye staring back at us!

METEOR SHOWERS

Surprisingly for such a small constellation, Coma Berenices does have a meteor shower associated with it. The Coma Berenicids are active from 8 December to 23 January, but with just 5–10 meteors per hour, this is not one of the more prominent showers.

MYTHOLOGIES

Soon after Berenice II took the throne of Egypt, her husband, co-ruler and cousin, Ptolemy III Euergetes, went off to war. It's said that Berenice was so worried for his safety that she offered to cut off her long, beautiful hair in sacrifice to the goddess Aphrodite for the safe return of her husband. When he returned safely, she dutifully cut off her tresses and placed them in Aphrodite's temple.

However, the hair Berenice had cut off then disappeared from the temple. Ptolemy was understandably furious, and things looked grim for the priests who ran the temple; they should have taken better care of the queen's holy offering. But then fast-thinking court astronomer Conon of Samos revealed that he had found Berenice's hair: the goddess had been so pleased with the sacrifice that she had placed it in the night sky. This was good news for all concerned – but especially for the priests.

INTERESTING FACTS

Conon's constellation wasn't accepted by everyone. The astronomer Eratosthenes of Cyrene (276–194 BCE), who lived in the region at the time, referred to these stars both as Berenice's hair and also as the hair of the mythical Ariadne. When the Greek-Egyptian astronomer Claudius Ptolemy wrote his *Almagest* in about 150 CE, he did not include Coma Berenices in his list of 48 constellations.

It seems that European astronomers rediscovered Conon's constellation. German cartographer Caspar Vopel included it as a separate constellation on a celestial globe in 1536. Danish astronomer Tycho Brahe is sometimes credited with 'creating' this constellation; in fact, he included what was already a well-established pattern in his star catalogue of 1602, but that star catalogue proved hugely influential, and Coma Berenices has appeared on star maps ever since.

26. CORONA AUSTRALIS, THE 'SOUTHERN CROWN'

Pronounced: 'koh-ROE-nuh oh-STRAHL-iss'
Short: CrA
Brightest star: α Coronae Australis or Meridiana (RA 19h 9m, Dec. −37°54')

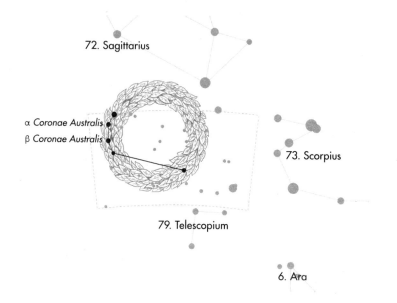

This is one of the smallest constellations, the 80th of 88 in terms of size, but it has been recognised since ancient times. Greek poet Aratus (c. 315–240 BCE) wrote of two wreaths in the night sky, and in Ptolemy's Almagest, written about 150 CE, these are named the 'southern wreath' and 'northern wreath' (that is, **27. Corona Borealis**). Originally consisting of 13 stars, Ptolemy reallocated one of these to the neighbouring constellation **79. Telescopium**.

WHEN AND WHERE TO OBSERVE

Corona Australis is best seen in southern skies between May and July. In the northern hemisphere, it is visible from latitudes south of +40°, such as the southern Mediterranean, in August. It is located between the constellations of **72. Sagittarius** and **73. Scorpius**, with **79. Telescopium** to the south and **6. Ara** to the southwest.

THE BRIGHTEST STARS

α Coronae Australis is a white A-type star some 2.6 times the mass and diameter of the Sun. It's known as Meridiana, from the Latin for 'south'

– an abbreviated version of Alphecca Meridiana, meaning 'the southern [version of] Alphecca', the brightest star in **27. Corona Borealis**.

β Coronae Australis is an orange K-type star that, having exhausted its supply of hydrogen, has expanded to some 43 times the diameter of the Sun, with about five times the mass. It's actually 615 times as luminous as the Sun, placing it among the brightest K-type stars visible to the naked eye.

OTHER BODIES

This constellation does not contain any Messier objects, but the Corona Australis Molecular Cloud is the closest region of star formation to us at just 430 light years away, and it generates low-mass stars. There are three distinct regions in the cloud: NGC 6726, NGC 6727 and NGC 6729, but our view of the heart of the cloud is obscured by great bodies of dust.

METEOR SHOWERS

There is one main meteor shower associated with this region. The Corona Australids occur between 14 and 18 March each year, but they usually have a low rate of around five meteors per hour. In 1992 astronomers detected rates of 45 per hour; we think that during this period our planet travelled through a more dense region of the debris field.

MYTHOLOGIES

To some Greek and Roman writers, the southern wreath – or crown – had fallen from the head of nearby **72. Sagittarius**. Others thought it had fallen much further and originally belonged to **19. Centaurus**.

In Chinese astronomy, the stars of this constellation sit within the northern quadrant of the sky so are part of the Black Tortoise of the North.

The Australian aboriginal Boorong people of northwestern Victoria saw this group of stars as Won, a boomerang thrown by Totyarguil, otherwise known as the star Altair, which sits in the constellation **5. Aquila**.

INTERESTING FACTS

Surviving coins from the city states of ancient Greece often depict the head of a god wearing a wreath of leaves and flowers. Each god had their own dedicated plant: for example, Aphrodite wore a wreath of myrtle, Apollo a wreath of laurel and Zeus a wreath of oak. Wreaths were also used to decorate stone statues and were worn by humans on special occasions such as at festivals and in ceremonies; we think this was a way to link the wearer to specific gods.

By Roman times, wreaths were being awarded to soldiers for particularly brave service. Victorious generals, such as the famous Julius Caesar (100–44 BCE), were awarded laurel wreaths. We sometimes speak of such people being 'crowned' with a wreath, as crowns are another piece of headgear used to symbolise authority and power.

27. CORONA BOREALIS, THE 'NORTHERN CROWN'

Pronounced: 'koh-ROE-nuh bor-ee-AHL-iss'
Short: CrB
Brightest star: α Coronae Borealis or Alphecca (RA 15h 34m, Dec. +26°42')

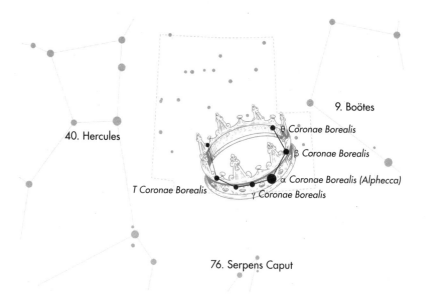

This is one of the 48 constellations listed by Greek-Egyptian astronomer Claudius Ptolemy in his Almagest, written about 150 CE. The constellation is bordered by **9. Boötes** to the north and west, **76. Serpens Caput** to the south and **40. Hercules** to the east.

WHEN AND WHERE TO OBSERVE

This northern constellation is wholly visible to observers north of latitude −50° and is best seen during July.

THE BRIGHTEST STARS

α Coronae Borealis is a binary star comprised of a blue-white A-type and a yellow G-type. The latter is a little smaller and less massive than the Sun, but it is hotter and gives off more radiation in the X-ray part of the spectrum: evidence that it is much younger than the Sun.

The Arabic name Alphecca is a shortened version of a phrase meaning 'the brightest star in a broken ring', the suggestion being that the constellation is not a wreath but a necklace or crown that has come apart.

β Coronae Borealis, γ Coronae Borealis and θ Coronae Borealis are all binary stars, too.

OTHER BODIES

With an apparent magnitude of about 10, T Coronae Borealis is not normally visible to the naked eye, but it has been known to flare very brightly on two occasions, in 1866 and 1946, when it briefly had an apparent magnitude of 2.5, earning it the nickname 'Blaze Star'.

This is another binary star; its two components orbit one another at a distance of about half that between Earth and the Sun. One is a cool red giant, which is losing material to its counterpart, a hotter white dwarf. The white dwarf is surrounded by a disc of this material, and both stars are partly obscured inside a cloud of it. The blazing flare-ups are the result of huge explosions caused by nuclear activity in the red dwarf.

METEOR SHOWERS

There are two main meteor showers associated with this constellation. The Northern Crown meteor shower occurs in late April or early May, while the Corona Borealis meteor shower, occurs in late September or early October.

MYTHOLOGIES

The Greeks thought this pattern of stars was a wreath of plants and flowers rather than a crown (see **26. Corona Australis** for more about the difference), and that this particular wreath had been given by the god Dionysus to Ariadne, princess of Crete.

According to legend, Ariadne's father King Minos ruled over the other Greek city states of the time and required them to send young men and women to Crete once a year. Theseus, the brave prince of Athens, was bothered that these young people – including his friends – were then never seen again. One year, Theseus volunteered to be one of the 'tributes' sent to Crete, so that he could find out what was really happening.

He duly discovered that the young people were being sent into a dark, twisting labyrinth under Minos' palace, where they were killed and eaten by the monstrous Minotaur. Theseus was also sent into the labyrinth, but Ariadne – who had fallen in love with Theseus – managed to provide him with a sword and something else to help him find his way in the darkness.

In one version of the story, she gave Theseus the wreath she'd already received from Dionysus; its god-given power made it shine bright in the dark. In another account, Ariadne instead gave Theseus a ball of string. Having defeated the Minotaur, Theseus abandoned Ariadne and, heartbroken, she ended up marrying Dionysus instead, who gave her a wreath for their wedding.

(In another version of the story, Ariadne received the wreath or crown as a wedding present from the goddess Aphrodite; in yet another, Theseus didn't get the wreath from Ariadne at all, but took it from a sea nymph called Thetis.)

In Celtic mythology, these stars represent Lady Arianrhod, goddess of fertility, rebirth and the weaving of cosmic time and fate. Her name has been translated as 'silver wheel', a symbol that represents the ever-turning wheel of the year, which ties in nicely with the shape of this constellation.

INTERESTING FACTS

Chinese astronomers also saw this pattern of stars as a loop, but to them it was Guansuo, a prison for criminals from the working classes. Just as this 'northern crown' has its southern counterpart, the Chinese prison was paired with another constellation, Tianlao, a prison for the upper classes, found in **83. Ursa Major.**

28. CORVUS, THE 'CROW'

Pronounced: 'KOR-vus'
Short: Crv
Brightest star: γ Corvi or Gienah (RA 12h 15m, Dec. −17°32')

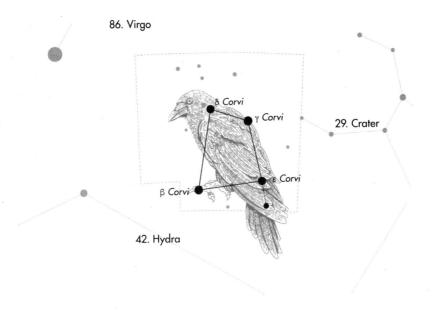

*The Babylonians knew this pattern of stars as MUL.UGA.MUSHEN, the 'raven', at least as far back as 1000 BCE. They saw this raven as standing on the back of a snake (that is, **42. Hydra**), and it's thought these constellations were associated with the rains of autumn, or perhaps even winter and death. It may be that the Greeks inherited this interpretation.*

WHEN AND WHERE TO OBSERVE

This constellation is located in the southern sky, but it is visible in the northern hemisphere from January to May. It can be seen at latitudes +60 to −90°, which includes all of the UK.

THE BRIGHTEST STARS

The brightest star, γ Corvi, is also known as Gienah, from an Arabic phrase meaning 'right wing of the crow', though it's often depicted in modern star charts as the left wing. It's a binary star comprising a giant, blue-white B-type and a smaller, dimmer K-type or M-type companion. The two stars orbit one another at a distance of about 50 times that between Earth and the Sun – a little further than the distance between the Sun and Pluto!

OTHER BODIES

As with many of these smaller constellations, there are no Messier objects here. However, the constellation contains numerous galaxies and a planetary nebula that can be seen with non-professional equipment. The planetary nebula NGC 4361 is located in the heart of Corvus.

METEOR SHOWERS

The recently discovered Eta Corvids were first recorded in January 2013. They were observed producing a total of 300 meteors between 20 and 26 January.

MYTHOLOGIES

By about 500 BCE, it seems that Corvus and adjacent constellations **29. Crater** and **42. Hydra** were identified with a particular legend in which the god Apollo sent a crow to fetch a cup of fresh water from a spring.

The greedy crow was distracted by an abundant fig tree nearby and decided to wait for the fruit to ripen. After several days, the fruit was ready and the bird ate its fill. Then it had to explain to Apollo why it had taken so long to return with his water.

The crow told Apollo that it had been attacked by a water snake that jealously guarded the spring. As 'proof' of this, it presented him with an entirely innocent water snake it had caught near the spring. Apollo easily saw through the lie and as punishment placed crow, cup and snake in the night sky, the cup forever just out of reach of the greedy bird.

The Tucano people of the northwest Amazon also recognise this pattern of stars as a bird associated with water: in this case, a kind of finely plumed heron called an egret. They see another egret in the constellation **25. Coma Berenices**.

INTERESTING FACTS

The Sail, also known as Spica's Spanker, is an asterism formed by the four brightest stars in the constellation. Two of these stars point in the direction of Spica, the brightest star in the nearby constellation **86. Virgo**. Sailors, who used the stars for navigation, saw in β Corvi, δ Corvi, γ Corvi and ε Corvi the corners of the distinctly shaped 'spanker' sail used on square-rigged ships, hence the name given to these four stars. We can still use bright Spica to find these four stars, and Corvus.

29. CRATER, THE 'CUP'

Pronounced: 'KRAY-ter'
Short: Crt
Brightest star: δ Crateris or Labrum (RA 11h 19m, Dec. −14°46')

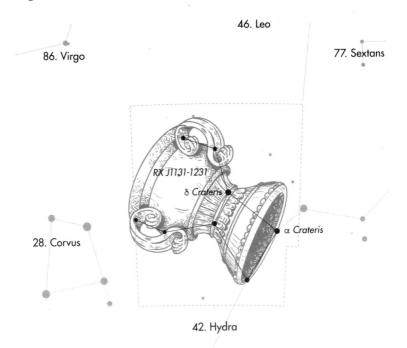

In ancient Greece, a krater was a large vase, usually made of clay, in which wine was mixed with water before being served to guests (who drank it from smaller, more manageable vessels). The earliest known examples of kraters date from the 600s BCE, and surviving texts from the period describe the etiquette involved in who did the mixing and how much the wine should be diluted.

WHEN AND WHERE TO OBSERVE
Best observed during April evenings, this constellation is wholly visible to observers south of +65° latitude.

THE BRIGHTEST STARS
The constellation's two brightest stars, δ Crateris and α Crateris, are both giant, orange K-type stars. The former is also known as Labrum, from the Latin for 'lip [of the cup]'.

OTHER BODIES

High-end telescopes have discovered a quasar known as RX J1131-1231 in this constellation. Quasars seem star-like when viewed from Earth, but they are very bright when viewed in the X-ray and radio part of the spectrum. This particular quasar is spinning extremely fast, at almost half the speed of light. We think this is because at its heart there is a supermassive black hole.

METEOR SHOWERS

Between 11 and 22 January there is a weak meteor shower called the Eta Craterids, peaking on 16/17 January.

MYTHOLOGIES

This pattern of stars was associated with a *krater*-type vase from at least as far back as about 500 BCE and linked to the story of a crow sent by the god Dionysus to gather water from a spring (see **28. Corvus**).

INTERESTING FACTS

Why does this constellation have the same name as the well-known circular impressions on the surface of the Moon and other bodies in space?

Over time, the word *krater* came to be used to describe smaller vessels – no longer large, heavy vases, but almost any bowl, goblet or small cup. By the 1600s, the word 'crater' was also used to describe the mouth of a volcano, usually a bowl- or funnel-shaped hollow that was the source of eruptions. The word was additionally used in a military sense to describe the hollows left behind by explosions.

German astronomer Johann Hieronymus Schröter (1745–1816) seems to have been the first to apply the word to the bowl-like impressions seen (by telescope) on the surface of the Moon. Debates raged over what had caused these lunar craters: were they signs of volcanic activity, the result of explosions or something else?

By sending probes and people to the Moon in the 1960s and 1970s, we gained conclusive evidence about the origins of these formations. The Moon's craters – and the craters we've found elsewhere – come from meteorites and asteroids crashing into the surface. We often refer to them as 'impact craters' to underline this.

30. CRUX, CONTAINING THE 'SOUTHERN CROSS'

Pronounced: 'KRUCKS'
Short: Cru
Brightest star: α Crucis or Acrux (RA 12h 26m, Dec. −63°05')

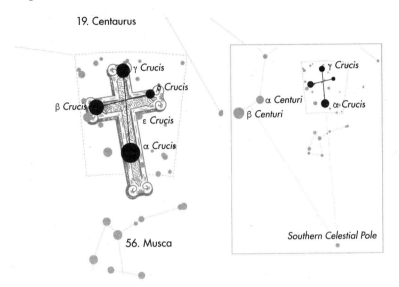

Although this is the smallest of the 88 constellations, Crux is easy to spot in the night sky. The four bright stars that form the asterism known as the Southern Cross comprise a kite shape or cross. On my first visit to Australia, this was the constellation I most wanted to see, as it is not visible from the UK. Fortunately, I was working out at the Woomera rocket range in South Australia, so I was able to get my first glimpse of it in clear skies, unpolluted by city lights.

WHEN AND WHERE TO OBSERVE

In the southern hemisphere, this constellation is easy to spot and it is visible throughout the year at latitudes south of −34°. In the northern hemisphere this constellation can be seen briefly at or below latitudes of +26°. For instance, if you are in Hawaii in May or June, you may spot Crux low on the horizon.

THE BRIGHTEST STARS

The brightest star, α Crucis or Acrux, sits nearly 323 light years away and is really a system of at least six interrelated components. The two brightest ones are B-type stars many times more massive and brighter than the Sun. β Crucis is a blue-coloured binary star, while γ Crucis is a binary with a red-coloured, M-class primary; at 88.6 light years away, this is the closest

M-class star to our Sun. δ Crucis is a blue-coloured B-class star. A fifth star, ε Crucis, is an orange K-type star.

OTHER BODIES

The Coalsack Nebula, known as a dark nebula, can be seen clearly with the naked eye and sits in the southern part of Crux, overlapping with the adjacent constellations of **19. Centaurus** and **56. Musca**. The Coalsack consists of clouds of dust and gas that obscure our view of our own galaxy, the Milky Way.

METEOR SHOWERS

Two very weak meteor showers have been mentioned in association with this constellation. The Alpha Crucids meteor shower reaches peak activity around the spring equinox, 21/22 March, while the Gamma Crucids — which were only discovered recently as a result of monitoring meteor activity in the area — reach their peak in mid-February. We don't currently know which comets are responsible for the dust that causes these low-activity meteor showers.

MYTHOLOGIES

When peering into the night sky from the southern hemisphere, we are actually looking deep into the heart of our own galaxy, the Milky Way. As a result, many more stars are visible from the southern hemisphere.

Given this wealth of stars, many aboriginal cultures created constellations not out of stars but from the clouds of interstellar dust that obscured them. The Coalsack Nebula is one such aboriginal constellation. The Aborigines also saw the Southern Cross as symbolising animist spirits that were integral to their ancestral beliefs. They saw Crux and the Coalsack Nebula as marking the head of the 'Emu in the Sky'.

There are a number of myths and legends about how the emu got into the sky, but one of the most popular is set during the early days of the world, in what is often known as the Dreaming. The story goes that a blind man and his wife lived in the bush. As the man was blind, he encouraged his wife to go out and find eggs for them to eat. However, whenever his wife returned with eggs, he would moan that they were too small.

One day, she found a large emu, which she thought would produce larger eggs. She threw rocks at it, hoping to distract it so that she could steal some of its eggs. However, the rock throwing angered the emu and it killed the woman.

Meanwhile, the husband was getting worried about his wife and ventured out to try to find her — as well as something to eat. As he searched, he stumbled upon some berries that restored his sight. Now that he could see, he set off in earnest to find his wife and found tracks leading to the large emu. On finding his wife's body, rather than killing the emu, he chased it into the sky, where it remains to this day.

INTERESTING FACTS

These stars have been known since ancient times, but Claudius Ptolemy considered them part of neighbouring **19. Centaurus**, rather than a constellation of their own.

In the 1500s, European sailors travelling to India and islands in the Indian Ocean made frequent reference to these stars, as they could be used in navigation. A line from γ Crucis to α Crucis points due south; in fact, continuing that line for 4.5 times the distance reaches the south celestial pole. That point also intersects with a line drawn at 90° to one between α and β Centauri, the so-called Southern Pointers.

Given its value in navigation, Crux was first recognised as a separate constellation in 1592 – although, due to inaccurate data, it was initially located in the wrong place in the sky. Better observations were made in Madagascar by Dutch sailors Pieter Dirkszoon Keyser, Frederick de Houtman and Vechter Willemsz over the winter of 1595–6 (see **22. Chamaeleon**), and these were used by Petrus Plancius (1552–1622) to establish Crux as its own constellation and in its proper place in 1598.

Crux was so integral to navigating the southern seas that Australia, Brazil, New Zealand, Papua New Guinea and Samoa include the constellation in their national flags. In some ways, they've become a symbol of southern identity.

Petrus Plancius

Born Pieter Platevoet in Dranouter in what is now Belgium, Plancius became a minister in the Dutch Reformed Church. In 1585, suddenly facing prosecution because of his faith, he fled to the Protestant Netherlands, but was short of funds.

Trading valuable goods and spices brought from India and the Indian Ocean was potentially profitable, but hazardous routes meant ships were often lost on the journey. Improved maps – of the sea, but also of the night sky – would mean ships had a better chance of following safe routes, increasing profits for all involved. Plancius's diligent work made him rich. Plancius didn't only make money by producing better maps. In 1602 he was a founder member of the Vereenigde Oost-Indische Compagnie (the Dutch East India Company).

But there was a darker side to all this commercial success. The Dutch East India Company had a long history of involvement in the slave trade, as did many other companies who made use of the improved maps and star charts. It has been argued that the vast international slave trade, which enslaved millions of people over centuries, wouldn't have been possible without this improved knowledge of the sea and stars.

31. CYGNUS, THE 'SWAN'

Pronounced: 'SIG-nuss'
Short: Cyg
Brightest star: α Cygni or Deneb (RA 20h 41m, Dec. +45°16')

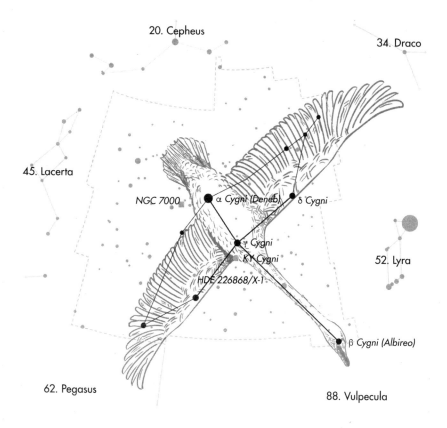

This distinctive and easily spotted pattern of stars is sometimes referred to as the Northern Cross, the counterpart to the Southern Cross (30. Crux).

WHEN AND WHERE TO OBSERVE

Visible between +90 and −40° latitude, this is one of the most easily recognisable constellations in the late summer and autumn sky of the northern hemisphere, with its distinctive cross shape and bright α Cygni, or Deneb. It's at its most prominent in September but is visible from June to October.

To see it, simply look above the eastern horizon just after sunset. Deneb is the first star you'll notice as the sky darkens; it sits at the top of the 't' shape of the constellation and is the tail feather of the swan.

THE BRIGHTEST STARS

α Cygni is a white A-type supergiant – one of the largest A-type stars we know of, some 20 times more massive than our Sun. It's also extremely bright, but estimates vary about the star's distance from Earth, so calculations of its luminosity range between 55,000 and 196,000 times that of the Sun. The name Deneb comes from the Arabic for 'tail [of the swan]'.

β Cygni is a binary star: a golden-yellow K-type star in notable contrast to a fainter blue B-type, both of which are visible when seen through a telescope. γ Cygni or Sadr (meaning 'chest') is a yellow F-type star, and δ Cygni is a triple star system with a prominent blue-white B-type and two fainter companions.

OTHER BODIES

In dark skies, the North America Nebula (NGC 7000) is visible to the naked eye. This is a good example of an emission nebula, which is emitting light because its hydrogen gas has been ionised by ultraviolet radiation from a very hot nearby star. However, the nebula also obscures the star, so it's not easy to see from Earth: we think it's an O-type with a temperature of more than 40,000K.

This constellation contains many more nebulae, open clusters and other deep-sky objects that can be seen with binoculars and telescopes.

The star HDE 226868 is also obscured from view by a huge disc of gas and dust, an 'accretion disc' of matter being pulled in by the enormous gravitational force of Cygnus X-1, the first black hole ever detected. The gravitational force increases the closer you get to a black hole, and at a certain point – known as the event horizon – even light cannot escape. The effect of this is that we can't see or detect black holes directly; we can only see their effects on matter outside the event horizon, such as the accretion disc.

KY Cygni is also not visible to the naked eye, but it is notable because, at some 1,033 times the diameter of the Sun, it is one of the largest stars we know of.

METEOR SHOWERS

The Kappa Cygnids occurs from 3 to 25 August every year, with the peak occurring around 17 August. However, even at peak intensity there are only around three meteors an hour. The Perseids meteor shower – emanating from **63. Perseus** – occurs during the same time period, with more frequent and brighter meteors, so the Kappa Cygnids are not as well known.

MYTHOLOGIES

To Chinese astronomers, these stars signified a bridge across the 'river' of the Milky Way. The ancient Greeks may have been influenced by the same idea of the Milky Way as a river, as they saw in this pattern of stars a bird associated with water; in the long tail of the cross, they saw the slender neck of a swan.

Cygnus (from the Greek for 'swan') is one of the 48 constellations listed by Greek-Egyptian Claudius Ptolemy in his *Almagest* of about 150 CE. Eudoxus of Knidos (c. 400–c. 350 BCE) and Greek poet Aratus (c. 315–240 BCE) also referred to the constellation, but their original works no longer survive and there's some debate about whether they and other writers of the period saw these stars as generally bird-shaped, rather than a swan specifically.

Whenever it was that the stars came to be associated with a swan, there were differing views about which swan it might be. To some, it represented the Greek god Zeus, who was said to have transformed himself into a swan to seduce the Spartan queen Leda (mother of the twins in **38. Gemini**). In other accounts, the gods turned a human man named Cygnus or Cycnus into a swan – but there are several different characters of that name, each with a different reason for having been transformed.

INTERESTING FACTS

Launched in 2009, the Kepler Space Telescope was initially focused on an area covering part of Cygnus and neighbouring constellations **34. Draco** and **52. Lyra** in its search for so-called exoplanets – planets in orbit round stars other than the Sun. To date, this mission has found a confirmed 2,710 exoplanets, as well as a number of candidates still to be confirmed.

Let's finish on something easy to spot with the naked eye. The Summer Triangle is visible in the northern hemisphere throughout the year, but in July and August it is almost directly overhead at midnight. It's made up of the three brightest stars of three separate constellations: α Cygni or Deneb in Cygnus, α Aquilae or Altair in **5. Aquila** and, brightest of all, α Lyrae or Vega in **52. Lyra**. The Summer Triangle frames a notable portion of the Milky Way, which seems to split in two at α Cygni – though you'll need clear skies and low levels of light pollution to see this.

32. DELPHINUS, THE 'DOLPHIN'

Pronounced: 'del-FIN-us'
Short: Del
Brightest star: β Delphini or Rotanev (RA 20h 37m, Dec. +14°35')

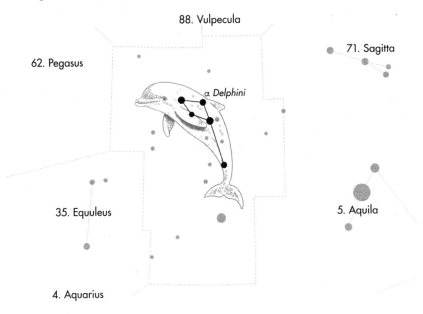

88. Vulpecula

71. Sagitta

62. Pegasus

α Delphini

35. Equuleus

5. Aquila

4. Aquarius

The dolphin constellation was one of the 48 listed by Greek-Egyptian astronomer Claudius Ptolemy in his Almagest of about 150 CE, but there's evidence that it was recognised long before then. The four brightest stars in the constellation form a slightly wonky diamond, an asterism known as 'Job's coffin' – but we don't know who Job was, or who first came up with this name!

WHEN AND WHERE TO OBSERVE

Delphinus is a small, faint constellation close to the celestial equator. It is visible from the northern hemisphere and best observed in late summer, around September. It can be observed at latitudes of 90 to −70°, so it makes a brief appearance in the southern hemisphere in late September and early October.

THE BRIGHTEST STARS

With the naked eye, β Delphini appears to be a white star; even through a telescope it is difficult to discern that this is really a binary system, the components of which are two F-type stars between 1.5 and 2 times the size of the Sun.

The slightly less bright α Delphini is also a binary star, the brighter component being a blue-white B-type star. Through a telescope, five relatively faint stars can be seen nearby, but we think (although at the moment we're not sure) that they only appear to be close when viewed from Earth and are not part of the same system.

OTHER BODIES

Delphinus contains no Messier objects, but due to its location with the Milky Way galaxy as a backdrop, there are a few prominent deep-sky objects contained within its boundaries. However, these objects are so faint that only large, professional telescopes can see them. They feature multiple globular star clusters, each of which have hundreds of thousands of stars, and a planetary nebula known as the Blue Flash Nebula (NGC 6905). A globular cluster called NGC 6934 is around 50,000 light years away from Earth.

METEOR SHOWERS

There are no known significant meteor showers associated with this constellation.

MYTHOLOGIES

According to legend dating back at least as far as the Greek historian Herodotus (c. 484–c. 425 BCE), the poet Arion of Lesbos was captured by pirates who decided to kill him. Arion persuaded the pirates to let him play one last song on his *kithara* – a seven-stringed lyre (see **52. Lyra**) – after which he threw himself into the sea. However, the brilliance of his playing had attracted a pod of dolphins, and one of these carried Arion to safety. For this, it was rewarded by being placed among the stars.

Greek-Egyptian astronomer Eratosthenes of Cyrene (276–194 BCE) gave a different account of the dolphin-shaped constellation. When the beautiful Amphitrite fled from the attentions of Poseidon, the Greek god of the sea, Delphinus was one of the minions sent out to find her. In one version of the story, Delphinus reported Amphitrite's whereabouts to Poseidon, who then captured her; in another, Delphinus persuaded Amphitrite to marry Poseidon and even took charge of the wedding. Poseidon rewarded this service by placing Delphinus among the stars.

INTERESTING FACTS

In 1814, the star catalogue produced by the Palermo Observatory in Sicily gave names to the stars α Delphini and β Delphini: Sualocin and Rotanev, respectively. It was only decades later than another astronomer worked out the origin of these names: backwards, they spell 'Nicolaus Venator', the Latinised version of Niccolò Cacciatore, assistant to Giuseppe Piazzi, who was director of the Palermo Observatory when the stars were named.

The names may have been part of an elaborate pun: perhaps Piazzi already knew that Cacciatore would succeed him as director of the observatory (which he did, from 1817 until he died in 1841); in France the heir apparent to the crown was traditionally known as the *dauphin* – meaning 'dolphin' – and, of course, this is the dolphin constellation. It's the sort of convoluted in-joke that might appeal to an old astronomer after many long, lonely nights of observing!

Alternatively, Cacciatore may have named the stars after himself, and his secret wasn't discovered until after his death. If this is the case, you have to admire his cheek!

33. DORADO, THE 'GOLDFISH'

Pronounced: 'doh-RAH-doe'
Short: Dor
Brightest star: α Doradus (RA 4h 33m, Dec. −55°02')

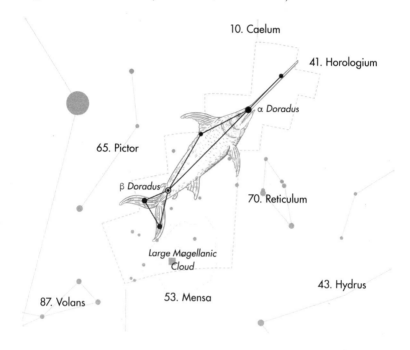

Compared to many of the great characters associated with constellations such as **40. Hercules** and **1. Andromeda**, and even some of the more mundane articles such as the compass in **23. Circinus**, a constellation called the 'Goldfish' seems very down to earth. But this constellation was originally meant to represent Coryphaena hippurus, better known as the mahi-mahi or common dolphinfish. These fish have nothing to do with dolphins, which are mammals, and they have impressive golden sides. In Spanish they are known as dorado, meaning 'gilded'. It seems that at some point this was mistranslated into 'goldfish'.

WHEN AND WHERE TO OBSERVE

This relatively small, faint constellation is best observed in southern skies. From November to January, it is visible from all latitudes south of +20°.

THE BRIGHTEST STARS

Blue-white α Doradus is one of the brightest binary stars visible, comprised of an A-type and a B-type completing orbits around one another every

12 years. β Doradus is a yellow Cepheid variable (see **20. Cepheus**), which undergoes regular fluctuations in brightness – changing from F-type to G-type – every 9.8 days.

OTHER BODIES
The most notable feature in (and just beyond) this constellation is the Large Magellanic Cloud, a small satellite galaxy to our own Milky Way, sitting at about 170,000 light years away. The earliest known reference to this is as one of three 'canopies' observed by the Italian explorer Amerigo Vespucci (1451–1512) – from whose name we get 'America'.

Some 15 years after Vespucci, the Large Magellanic Cloud was described as a 'cloud' by fellow Italian explorer Antonio Pigafetta (c. 1491–c. 1531), who was part of the Magellan expedition of 1519–22 that made the first trip right around the world (see **3. Apus**). After this, sailors seem to have referred to this and its smaller companion (found in **43. Hydrus** and **82. Tucana**) as the 'Magellanic clouds', but the name doesn't appear to have caught on with astronomers until as late as the 1840s.

METEOR SHOWERS
There are no significant meteor showers associated with this constellation.

MYTHOLOGIES
The Chinese astronomer Xu Guangqi (1562–1633) included the stars of Dorado in two of his southern asterisms, the Goldfish and the White Patches Attached (Jibái and Jny, respectively). When Xu Guangqi created the Chongzhen Almanac at the conclusion of the Ming Dynasty, he included 23 near-Antarctic constellations/asterisms that had been seen in Western star catalogues but were left out of ancient Chinese charts. In doing so, he provided maps of the stars between −40° and −90° declination, an area not visible from China.

INTERESTING FACTS
On a trip to Indonesia in the 1590s, Dutch sailors Pieter Dirkszoon Keyser and Frederick de Houtman (see **22. Chamaeleon**) are thought to have observed dorados in pursuit of flying fish and were thereby inspired to name this constellation and the one it 'follows', **87. Volans**. Dorado was first shown on a star globe in 1598 by Petrus Plancius (see **30. Crux**).

German astronomer Johannes Kepler (1571–1630), whose laws of planetary motion are still used to calculate the distance at which planets orbit their stars, proposed an alternative name for this constellation: Xiphias, meaning 'swordfish'. Despite Kepler's renown, the name didn't catch on, though some sources today refer to Dorado as meaning 'swordfish'.

34. DRACO, THE 'DRAGON'

Pronounced: 'DRAY-koe'
Short: Dra
Brightest star: γ Draconis or Eltanin (RA 17h 56m, Dec. +51°29')

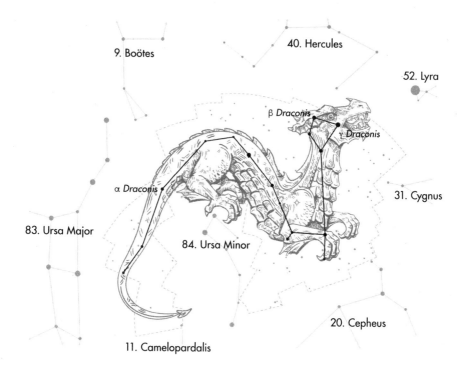

9. Boötes

40. Hercules

52. Lyra

β Draconis

γ Draconis

31. Cygnus

α Draconis

83. Ursa Major

84. Ursa Minor

20. Cepheus

11. Camelopardalis

This is the eighth largest constellation and yet it is not very prominent. It gets its name from the Latin word draconem, *which means 'great serpent' – an apt description for this constellation's snaking path across the northern sky. This is another of the 48 constellations listed by Greek-Egyptian astronomer Claudius Ptolemy in* Almagest *in about 150* CE, *apparently drawing from the now lost writings of Eudoxus of Knidos in the 300s* BCE.

WHEN AND WHERE TO OBSERVE
Close to the north celestial pole, Draco is visible all year round in the northern hemisphere and can be seen from anywhere north of −15° latitude. It's best observed in July.

THE BRIGHTEST STARS
Orange γ Draconis is a bright, K-type star known as Eltanin, from the Arabic for 'great serpent' or 'dragon'.

Blue-white α Draconis or Thuban, from the Arabic for 'large snake', is a binary star system that was, between 3942 and 1793 BCE, the northern pole star – that is, closer to the celestial north pole than any other star. Currently, the pole star is α Ursae Minoris (Polaris) in **84. Ursa Minor.**

Yellow β Draconis is also known as Rastaban, derived from the Arabic for 'head of the serpent'. It's a binary star, composed of a huge G-type some six times the mass, 40 times the diameter and almost 1000 times the brightness of the Sun, and a fainter companion. They orbit one another in a cycle lasting some 4,000 years.

OTHER BODIES

Draco is home to one Messier object. M102, also known as the Spindle Galaxy, is an edge-on galaxy with a dark dust lane running along its rim. It sits more than 40 million light years away from our Solar System.

This constellation also contains a few notable deep-sky objects: the beautiful Cat's Eye Nebula (NGC 6543), the Draco Dwarf Galaxy and the Tadpole Galaxy.

METEOR SHOWERS

The Draconids meteor shower usually occurs between 6 and 10 October, peaking on the night of 8/9 October. This shower is the result of Earth's motion through the dust and debris left by comet 21P/Giacobini–Zinner. It takes around 6.6 years for this comet to complete one orbit of the Sun. As with many showers, the number of meteors per hour can vary greatly depending on the density of the debris trail. Although it is unusual to see more than five meteors per hour in this shower, in 1933 and 1946 skywatchers witnessed thousands of meteors in a single hour – so it's definitely one to keep an eye on!

MYTHOLOGIES

The Greeks thought this constellation related to a great dragon called Ladon, who guarded a tree of golden apples. The 12 tasks imposed on heroic **40. Hercules** included stealing one of these apples, so he shot Ladon with a poisoned arrow. Alternatively, this constellation may have been a dragon thrown into the sky by the goddess Athene.

In Roman mythology, Draco was one of the giant Titans who fought the Olympian gods for ten years. The Titan was killed by the goddess Minerva and thrown into the sky, where it froze around the North Pole.

The name of this constellation in modern Chinese is 天龍座 (Tiānlóng Zuò), meaning 'the heaven dragon constellation'. In ancient Chinese astronomy, the stars of this constellation sat in two quadrants represented by the Black Tortoise of the North and the Three Enclosures.

INTERESTING FACTS

Observing γ Draconis in the 1720s, English astronomer James Bradley (1692–1762) made one of his two famous discoveries: an apparent movement in the position of the star on separate observations over one year was not due to the star being in motion but due to the movement of Earth. This discovery, known as the 'aberration of light', was further proof that Earth orbited the Sun and not the other way round. (Bradley later discovered that the orientation of the axis on which Earth spins is liable to change, a process known as nutation.)

35. EQUULEUS, THE 'LITTLE HORSE' OR 'FOAL'

Pronounced: 'eh-KWOO-lee-us'
Short: Equ
Brightest star: α Equulei or Kitalpha (RA 21h 15m, Dec. +5°14')

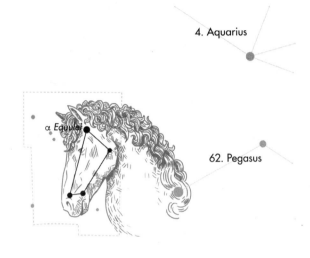

4. Aquarius

α Equulei

62. Pegasus

32. Delphinus

This horsehead-shaped pattern of stars is the second smallest constellation (after **30. Crux**). It is thought to have been devised by the Greek astronomer Hipparchus of Nicaea (c. 190–c. 120 BCE). It is sometimes known as Equus Primus, or the 'First Horse', because it rises just before the constellation **62. Pegasus**.

WHEN AND WHERE TO OBSERVE

Equuleus can be seen from most of the northern and southern hemispheres at latitudes ranging from +90 to –80°. This very small, faint constellation is best observed in northern skies in September. The easiest way to find it is by first spotting its more obvious neighbours, **4. Aquarius** and **62. Pegasus**.

THE BRIGHTEST STARS

The stars of this small constellation are all relatively faint. The brightest of them, α Equulei, is also known as Kitalpha, from an Arabic phrase meaning 'piece of the horse'. It's a binary star composed of a yellow G-type nine times the diameter and some 50 times brighter than the Sun, and a smaller, hotter A-type.

OTHER BODIES

Equuleus has no notable deep-sky objects due to its small size and distance from the plane of the Milky Way.

METEOR SHOWERS

There are no significant meteor showers associated with this constellation.

MYTHOLOGIES

Equuleus is associated with several Greek myths, most notably one about Hippe, the daughter of the centaur Chiron and the nymph Chariclo. Some claim that this constellation depicts a horse called Celeris (meaning 'swift'), the brother of winged **62. Pegasus** and property of Castor – one of the twins in **38. Gemini**. Other accounts suggest that it is the horse created by the god Poseidon during a contest, or the half-horse son of Saturn.

INTERESTING FACTS

The only surviving text by Hipparchus is his commentary on a poem by Aratus (c. 315–240 BCE), which was in turn based on a book about the stars and constellations by Eudoxus of Knidos (c. 400–c. 350 BCE). Roman writers seem to have had access to more works by Hipparchus. Pliny the Elder (c. 23–79 CE) says in his hugely influential *Natural History* – the earliest encyclopaedia to survive, and perhaps the first such book ever written – that Hipparchus was inspired to take up astronomy by the awesome sight of a supernova, and that he used various instruments to take very accurate measurements for his now lost star catalogue.

Some think Hipparchus used – perhaps even invented – the system of ranking apparent magnitudes of stars from 1 (bright) to 6 (faint). We still use a modified version of that system, extended beyond 6 to include stars observed by telescope, and into minus numbers, too. Our Sun is a star with an apparent magnitude of –26.7 (see 'Stellar Brightness: Real and Apparent', page 14).

To mark Hipparchus' contribution to astronomy, a European Space Agency satellite was named Hipparcos after him – the first spacecraft dedicated to precision astrometry (the accurate measurement of the positions of celestial objects). It was launched in 1989 and operated until 1993, during which time it catalogued nearly 120,000 stars.

36. ERIDANUS, THE 'RIVER'

Pronounced: 'eh-RID-ah-nus'
Short: Eri
Brightest star: α Eridani or Achernar (RA 1h 37m, Dec. −57°14')

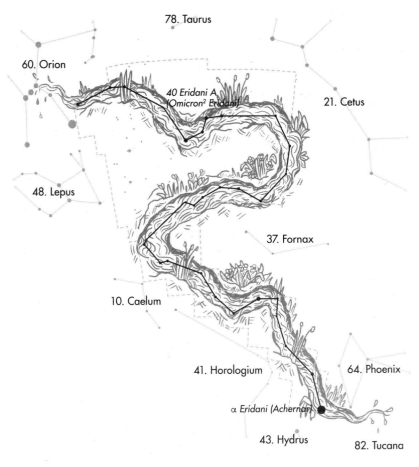

This is the sixth largest constellation and the one that stretches the furthest across the sky from north to south. Ancient cultures saw themselves in this long, meandering constellation. To the Babylonians it was the river Euphrates, to the Egyptians it was the river Nile – in both cases, the waters on which each civilisation depended. The Greeks and Romans knew the constellation as Eridanus, which was also the name they applied to the longest river in Italy, the one we know by the modern name Po. There's some debate about whether the river or the constellation had the name Eridanus first.

WHEN AND WHERE TO OBSERVE

Eridanus is located in the southern hemisphere and snakes across a huge part of the southern sky. From October to December it can be seen from latitudes south of +32°.

THE BRIGHTEST STARS

α Eridani is a binary star. The brighter component is a hot, blue B-type also known as Achenar, from the Arabic for 'end of the river'. Unusually fast rotation causes it to have an oblate or 'squashed' shape. It's in orbit with a smaller A-type.

40 Eridani, also known as Omicron^2 Eridani, is a triple star system just visible to the naked eye in dark skies. Its main component, 40 Eridani A, is also known as Keid, from the Arabic for 'eggshells', and is an orange K-type. Its companions are a D-type white dwarf and an M-type red dwarf.

OTHER BODIES

There are no Messier objects in Eridanus, but there are a few dozen faint galaxies and nebulae. These include the Witch Head Nebula, a reflection nebula, and the Eridanus Group of more than 200 galaxies.

Sadly not observable without professional equipment, the Eridanus Supervoid, discovered in 2007, is a large region devoid of galaxies. It is the second largest void of this kind we know of, with a diameter of about 1 billion light years.

METEOR SHOWERS

Eridanus is associated with two meteor showers. The Nu Eridanids are a newly discovered meteor shower that can be seen every year between 30 August and 12 September; their parent body is an unidentified Oort Cloud object. The Omicron Eridanids peak between 1 and 10 November.

MYTHOLOGIES

The name of the constellation comes from Greek mythology: when Phaethon, son of the sun god Helios, rode his father's flaming chariot too low in the sky, the whole Earth caught fire. The burning Phaethon threw himself into the river Eridanus. When the Argonauts later sailed up the same river, they found the remains of his body smouldering noxiously.

The Greek poet Aratus (c. 315–240 BCE) made the link between that mythological story and the constellation. However, when Greek-Egyptian astronomer Claudius Ptolemy included this as one of the 48 constellations in his *Almagest* of about 150 CE, he referred to it as Potamos – a generic 'river'. We're not sure why; perhaps he was aware that the Egyptians and Babylonians had also claimed it as their own.

INTERESTING FACTS

In 2018, a planet was discovered in orbit round 40 Eridani A, a super-Earth with a mass about 8.5 times that of our own planet and a year just over 42 days long. That means it must be very close to the star – closer than Mercury is to the Sun – and it must receive a high level of stellar radiation that would not be conducive to life.

Perhaps there are other planets yet to be discovered in the system. According to the science-fiction TV and film franchise *Star Trek*, another planet in orbit round 40 Eridani A is Ni'Var, or Vulcan, the home of Mr Spock.

37. FORNAX, THE 'FURNACE'

Pronounced: 'FOR-nacks'
Short: For
Brightest star: α Fornacis or Dalim (RA 3h 12m, Dec. −28°59')

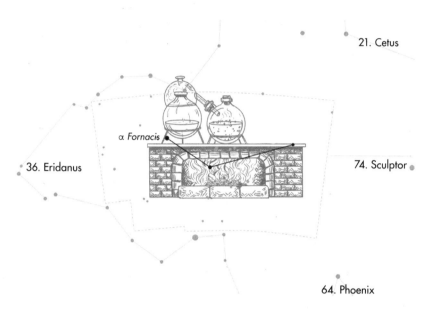

This midsized constellation is one of the 14 first named by Nicolas-Louis de Lacaille after his trip to the Cape of Good Hope to study the southern night sky (see **2. Antlia**). It is bordered on three sides by the snaking river **36. Eridanus**.

WHEN AND WHERE TO OBSERVE
Located in the southern hemisphere, Fornax can be seen at latitudes between +50° and −90° during the months around December.

THE BRIGHTEST STARS
Even the constellation's brightest star is relatively faint. α Fornacis is a binary system, its main component an F-type also known as Dalim. Oddly, this name was first applied by Italian astronomer Giuseppe Piazzi (1746–1826), from the Arabic al-ẓalīm, meaning 'ostrich'. Yet Arabic astronomers had already applied that name to other stars – α Eridani in nearby **36. Eridanus** and α Piscis Austrini (better known today as Fomalhaut) in **67. Piscis Austrinus**.

The other star in this binary system is an example of a so-called blue straggler. Bluer and brighter than other stars in the same cluster that we think are of the same age, it 'straggles' away from them when they are plotted on a graph of brightness against star class. We think this extra brightness is the result of two stars merging.

OTHER BODIES

Fornax does not have any Messier objects, but it does have a globular cluster, NGC 1049.

The Fornax Cluster contains 58 individual galaxies, the second richest such cluster within 100 million light years of Earth. Within the Fornax Cluster, NGC 1316 is a prominently bright elliptical galaxy; it is also one of the brightest radio sources in the sky.

Fornax also has a one-of-a-kind galaxy known as UDFy-38135539. This is extremely faint and the second most distant object ever observed in the universe. It is thought to be more than 13 billion light years away from Earth.

METEOR SHOWERS

There are no significant meteor showers associated with this constellation.

MYTHOLOGIES

Due to the late identification of this constellation, there are no significant mythologies associated with it. Coincidentally, Fornax was the goddess of bread and baking in Roman mythology, but this is not thought to have anything do with the name of the constellation.

INTERESTING FACTS

A furnace is, essentially, a chamber in which fuel can be burnt for the purpose of intensely heating something else. Historically, the word could be used to mean any kind of enclosed fireplace, oven, cauldron, etc., and it also described any place of extreme heat, such as a volcano or desert.

Lacaille initially called his constellation the 'chemical furnace' after an apparatus used in scientific experiments. In this pattern of stars, he saw such an apparatus, complete with a mechanism known as an ambic used in the distillation of liquids. At the time, this was modern, sophisticated kit.

In fact, Lacaille is thought to have named the constellation in honour of his good friend Antoine Lavoisier, the father of modern chemistry and a French scientist who was guillotined in 1794 during the French Revolution. The Latin version of Lacaille's name, *Fornax Chimiae*, was later shorted to Fornax.

38. GEMINI, THE 'TWINS'

Pronounced: 'JEM-uh-nye'
Short: Gem
Brightest star: β Geminorum or Pollux (RA 7h 45m, Dec. +28°1')

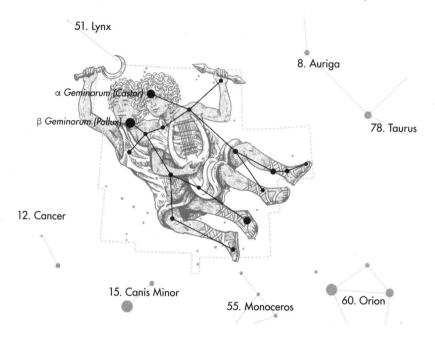

This medium-sized constellation is known to many as it is one of the signs of the zodiac, and because its two brightest stars are so prominent. Gemini was one of the 48 constellations catalogued by the Greek astronomer Ptolemy in the second century.

WHEN AND WHERE TO OBSERVE
The two bright 'twins' make this is a very distinctive constellation in the winter skies of the northern hemisphere, visible all through the night in December and January. The twins are easily located off the shoulder of another distinctive winter constellation, **60. Orion**.

THE BRIGHTEST STARS
β Geminorum or Pollux is in fact slightly brighter than α Geminorum or Castor, but these bright stars respectively mark the head of each twin. Despite the stories of the brothers' closeness, the stars only appear to be close together as seen from Earth; Pollux is 34 light years away from us, but Castor is another 17 light years beyond that.

Castor turns out to be system of six interlinked blue-white A-type stars, while Pollux is a giant, orange K-type. In 1993, evidence suggested a planet in orbit round Pollux, which was confirmed in 2006. This planet is at least twice the size of Jupiter and orbits Pollux every 590 days. Since 2014, the planet has been known as Thestias, another name for the twins' mother, Leda.

OTHER BODIES

Gemini contains one Messier object, which is easy to see with binoculars. This is an open cluster of about 100 stars, known as M35.

There are also a few interesting deep-sky objects here, but you will generally need a large telescope to see them. The Eskimo Nebula (NGC 2392) looks like a face within a fur parka. The Jellyfish Nebula is the supernova remnant of a star that exploded between 3,000 and 30,000 years ago. There is also the Medusa Nebula, a planetary nebula with glowing gas filaments that look like the snakes in the hair of the gorgon Medusa from Greek mythology.

METEOR SHOWERS

The Geminids meteor shower occurs between 19 November and 24 December every year, generally peaking on the nights of 13 and 14 December. Unlike most other meteor showers, the Geminids are the product of an asteroid rather than a comet. They're also one of the more impressive showers, with observers seeing up to 150 meteors per hour at peak times.

MYTHOLOGIES

To the Babylonians, this pattern of stars represented identical twins Lugal-irra and Meslamta-ea, guardians of the gate to the underworld who were supposed to chop up the dead before they could pass through. Depictions of them from the time seem always to show Lugal-irra on the left and Meslamta-ea on the right.

The Greek poet Aratus (c. 315–240 BCE) also referred to these stars as the twins, suggesting an inheritance from Babylon. Roman author Hyginus (c. 64 BCE–17 CE) described them not as twins but as half-brothers Apollo and **40. Hercules** from Greek mythology. Confusingly, when in about 150 CE Claudius Ptolemy listed this as one of his 48 constellations in *Almagest*, he referred to them as the twins, but in another work he linked them to Apollo and Hercules, too.

By this point, Greek-Egyptian astronomer Eratosthenes of Cyrene (276–194 BCE) had already linked this pattern of stars to other mythological brothers. Castor and Pollux were the sons of Leda, queen of Sparta. In some accounts they are twins, born after their mother was seduced by the god Zeus disguised as a swan (see **31. Cygnus**). In other versions, Zeus was father to Pollux and the famous Helen of Troy, and Castor was their half-brother.

Castor and Pollux were also seen as brothers by many aboriginal groups. For the Wergaia people of western Victoria, Australia, the stars represent the hunters Yuree and Wanjel who hunt and kill the kangaroo Purra.

In later life, Castor and Pollux were in the crew of the *Argo* on its various adventures (see **17. Carina**) and saved the ship from a terrible storm. It was later said that they had god-given powers to save imperilled sailors on this and other voyages, so this easy-to-spot constellation may have been seen as a symbol of good fortune by those sailors who used stars to navigate.

In Chinese astronomy, the stars in the constellation Gemini are located in two areas: the White Tiger of the West (西方白虎, Xī Fāng Bái Hǔ) and the Vermilion Bird of the South (南方朱雀, Nán Fāng Zhū Què).

INTERESTING FACTS

I've been privileged to work at the Gemini Observatory, twin telescopes located at different sites – the one in Hawaii covering the northern skies, the one in Chile looking south. They're among the largest telescopes of their kind in the world, each one fitted with an 8.1m reflecting mirror, and both at high altitudes to minimise the effects of Earth's atmosphere on what they can observe. The result is clear, detailed views of almost the whole sky (they can't quite see to the celestial poles).

39. GRUS, THE 'CRANE'

Pronounced: 'GRUSS'
Short: Gru
Brightest star: α Gruis or Alnair (RA 22h 8m, Dec. −46°57')

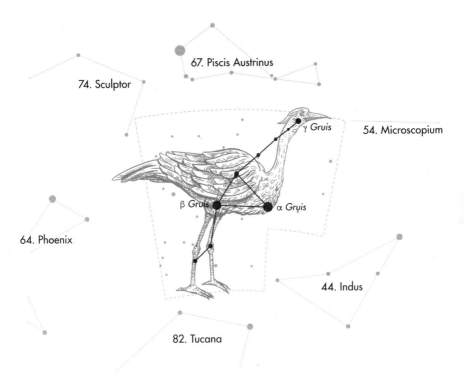

These stars were originally depicted as part of the tail of neighbouring fish **67. Piscis Austrinus**. Then, in 1598, Petrus Plancius (see **30. Crux**) and Jodocus Hondius produced a celestial globe based on observations made earlier that decade by Dutch explorers Pieter Dirkszoon Keyser and Frederick de Houtman (see **22. Chamaeleon**). On the globe, these stars were identified as a distinct constellation, Grus. In 1603, Johann Bayer's Uranometria was the first star atlas to include this constellation. In the early seventeenth century, it was briefly known as Phoenicopterus, which means 'flamingo' in Latin. It's now one of the Southern Birds, along with **61. Pavo, 64. Phoenix** and **82. Tucana**.

WHEN AND WHERE TO OBSERVE
Grus is located in the southern hemisphere. From July to September, it can be seen at latitudes south of +33°.

THE BRIGHTEST STARS

Sadly, the constellation is less notable than the distinctive birds it is named after. Blue-white α Gruis is a B-type star also known as Alnair, from an Arabic phrase meaning the 'bright one in the fish's tail', from when it was seen as part of **67. Piscis Austrinus**. Another blue-white B-type, γ Gruis, is known as Aldhanab, from the Arabic for 'tail'.

By contrast, β Gruis is an M-type red giant. It is known as Tiaki, a traditional name used on the Tuamotu Islands in the southern Pacific Ocean and thought to mean 'bright'.

OTHER BODIES

Grus does not contain any Messier objects. However, there are some interesting bodies associated with this constellation. IC 5148, known as the Spare Tyre Nebula, is a planetary nebula with an expansion rate of 50km/s, one of the fastest-expanding planetary nebulae known. It sits approximately 3,000 light years away from Earth.

NGC 7424 is a barred spiral galaxy about 100,000 light years across, similar in size to our own Milky Way.

METEOR SHOWERS

There are no significant meteor showers associated with this constellation.

MYTHOLOGIES

Grus is the Latin for 'crane'. The ancient Egyptians thought cranes flew very high and so used them as the symbol for astronomers. However, in *Star Tales*, Ian Ridpath suggests that Keyser and de Houtman had in mind the much larger sarus crane (*Grus antigone*) found throughout southeast Asia, which they would have seen on the voyage on which they recorded these southern stars.

INTERESTING FACTS

The common crane (*Grus grus*) is the tallest bird native to the UK, distinctive for its long legs, ruffled tail feathers and strange 'bugling' call. The famous 'dance of the cranes' involves pairs of birds reinforcing bonds with one another by stamping their feet, fluffing up their tail feathers, flapping their wings and throwing their heads back. That wild dance might help you remember this constellation!

40. HERCULES

Pronounced: 'HER-kyoo-leez'
Short: Her
Brightest star: β Herculis or Kornephoros or Rutilicus (RA 16h 30m, Dec. +21°29')

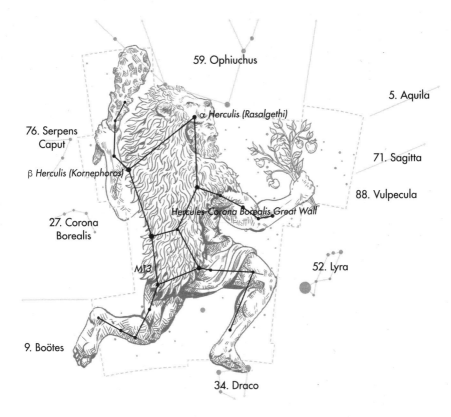

This is the fifth largest constellation in the sky, but its faint stars mean it can be difficult to find. It is best to find it by looking in the gap between brighter constellations **34. Draco** and **59. Ophiuchus**. Hercules also contains a distinctive asterism, the Keystone: a cluster of four relatively brighter stars that form a square, which makes up the hero's torso in the larger constellation.

WHEN AND WHERE TO OBSERVE
This constellation is best seen in the northern hemisphere during the summer. It is also visible in the southern hemisphere from May to August at latitudes ranging from +90 to −50°.

THE BRIGHTEST STARS

Yellow β Herculis is a binary star comprised of a G-type almost three times the mass of the Sun but 17 times the diameter, and a companion with a mass of about 90% that of the Sun. The names for this star suggest confusion about exactly how to depict Hercules: Kornephoros comes from the Greek for 'club bearer', while Rutilicus derives from the Latin for 'armpit'.

The distinctly red α Herculis is also known as Rasalgethi, from the Arabic for 'head of the kneeler' (see 'Mythologies', below). This is a multiple star system, and its two brightest components are orange and blue-green, respectively. Although the stars look close together, they're actually 500 times the distance from Earth to the Sun apart and take 3,600 years to complete an orbit of one another.

OTHER BODIES

Hercules contains two Messier objects: M13 (NGC 6205) and M92 (NGC 6341) are both globular clusters. M13, also known as the Hercules Cluster, is bright enough to be seen with the naked eye as a fuzzy patch; much more detail can be seen with a telescope. It's thought that this cluster contains around 300,000 stars. In 1974, a radio message with information about Earth and humanity was beamed in the direction of M13 from the Arecibo Observatory in Puerto Rico – a message it is hoped intelligent aliens will be able to decode – but it will take some 23,000 years for this message to arrive!

METEOR SHOWERS

The Tau Herculids meteor shower was discovered in 1930. The parent of this shower was 73P/Schwassmann–Wachmann, a periodic comet with a 5.4-year orbital period. However, the comet exploded in late 1995. Now, when Earth passes through the debris trail between 19 May and 19 June each year, the display can be as intense as 1,000 meteors per hour – or as weak as nothing at all.

MYTHOLOGIES

The Greek poet Aratus (c. 315–240 BCE) wrote that this pattern of stars was known as 'On-His-Knees', an unknown figure bent at some unknown task. By the time of historian Dionysius of Halicarnassus (c. 60 BCE– c. 7 BCE), this kneeler was the mythological hero Heracles, praying to his father – the god Zeus – for help with battling two giants. Greek-Egyptian astronomer Claudius Ptolemy included Heracles among the 48 constellations of his *Almagest* in about 150 CE. We now know the constellation by the Latin version of that name.

Depictions of Hercules have tended to show him in more heroic poses than humble prayer: battling **42. Hydra**, standing (or kneeling) on the snakes **34. Draco** or **76. Serpens** and/or raising a club or similar weapon over his head.

INTERESTING FACTS

In 2013, a team of American and Hungarian astronomers studying 283 gamma-ray bursts discovered that 19 of them were grouped together. What's now called the Hercules–Corona Borealis Great Wall is the largest known structure in the universe, measuring some 10 billion light years from end to end – more than 10% of the diameter of the visible universe (93 billion light years). The Great Wall is centred on the border between Hercules and **34. Draco**, but also spans **9. Boötes, 27. Corona Borealis** and **52. Lyra**.

The team that discovered this remarkable structure didn't give it a name. However, news of the discovery so impressed Filipino teenager Johndric Valdez that he created a Wikipedia page for it, which required a title. The name Valdez came up with – the Hercules–Corona Borealis Great Wall – quickly spread among astronomers.

41. HOROLOGIUM, THE 'CLOCK'

Pronounced: 'hor-uh-LOE-jee-um'
Short: Hor
Brightest star: α Horologii (RA 4h 14m, Dec. −42°17')

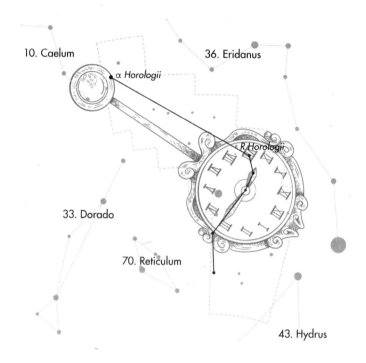

This is a small, faint constellation, and another of those created by French astronomer Nicolas-Louis de Lacaille in the eighteenth century (see **2. Antlia**). Lacaille originally called the constellation Horologium Oscillitorium, or 'the Pendulum Clock', but the name was later shortened.

WHEN AND WHERE TO OBSERVE
Horologium is located in the southern hemisphere of the sky. It is visible from latitudes south of +23° from October to December.

THE BRIGHTEST STARS
α Horologii is a relatively old, orange K-class star, some 1.55 times the mass of the Sun but 11 times its diameter, having burnt through all of its hydrogen fuel and then swollen up into a giant.

R Horologii is a red giant with one of the greatest ranges in apparent brightness of any star visible to the naked eye. Over a regular period of 404.8 days, it ranges between apparent brightness of 4.7 and 14.3!

OTHER BODIES

This constellation contains no Messier objects, but there are a number of deep-sky objects, mostly star clusters and galaxies. It also features the Horologium Supercluster, which contains more than 5,000 galaxy groups and more than 300,000 individual galaxies. This massive cluster covers an area of space approximately 550 million light years across.

METEOR SHOWERS

There are no significant meteor showers associated with this constellation.

MYTHOLOGIES

Due to its modern conception, and because the stars are relatively dim, there is little mythology associated with this constellation. In Chinese astronomy, this constellation, located within the western quadrant of the sky, symbolises the White Tiger of the West.

INTERESTING FACTS

Lacaille originally named this constellation in his native French as l'Horloge à pendule et à secondes, or 'a clock with a pendulum and a second hand'. That meant he had in mind the pendulum clock invented by Dutch scientist Christian Huygens in 1656, which made timekeeping much more accurate.

Lacaille died in 1762, and the following year a catalogue and star chart were published with the constellation now named in Latin as Horologium.

42. HYDRA, THE 'FEMALE WATER SNAKE'

Pronounced: 'HIGH-drah'
Short: Hya
Brightest star: α Hydrae or Alphard (RA 9h 27m, Dec. −8°39')

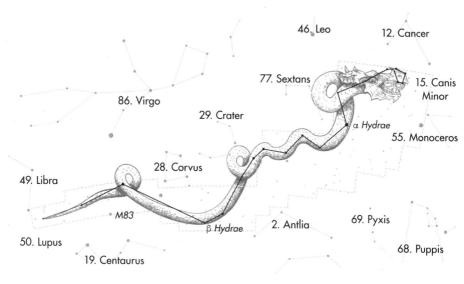

The largest of the 88 constellations is also the longest. Hydra borders 14 other constellations – more than any other. The Babylonians saw this pattern of stars and **76. Serpens** as serpents, and these interpretations were inherited and adapted by the Greeks. Greek-Egyptian astronomer Claudius Ptolemy included Hydra among the 48 constellations listed in his Almagest of about 150 CE.

WHEN AND WHERE TO OBSERVE
Running close to the celestial equator, this constellation is best seen from the southern hemisphere, but it can also be seen in the northern hemisphere between January and May. It can be seen at latitudes ranging from +54 to −83°.

THE BRIGHTEST STARS
Orange α Hydrae is a giant K-type star known as Alphard, from the Arabic for 'individual' or 'solitary', given that there are no bright stars around it.

The next brightest star in the constellation is blue-white β Hydrae, which is actually two stars that look close together as viewed from Earth.

OTHER BODIES

Three Messier objects can be found in this constellation. M48 (NGC 2548) is an open star cluster with about 80 stars and it can be seen with the naked eye. M68 (NGC 4590) is a globular cluster of more than 2,000 stars. M83 (NGC 5236) is better known as the Southern Pinwheel Galaxy and was also discovered by Lacaille, in 1752. Visible through binoculars, it is a magnificent, relatively close and bright example of a barred spiral galaxy.

METEOR SHOWERS

The Sigma Hydrids run from 3 to 15 December, peaking on 6 December; they are a very active shower with an unknown parent body. The Alpha Hydrids are a minor shower that peaks between 1 and 7 January.

MYTHOLOGIES

In Hindu astrology, this pattern of stars is seen as a coiled serpent also known as Nāga or the 'Clinging Star'. In the Greek tradition, it's a sprawling, winding snake. In some accounts, this was the snake caught by **28. Corvus** to explain why it took that wily crow so long to fill the cup **29. Crater** with water for the god Apollo.

The Hydra is also a snake-like monster associated with the legend of **40. Hercules**, dating as far back as Hesiod's *Theogony* of c. 730–700 BCE. According to this story, the Hydra dwelt in a lake at Lerna in southeastern Greece. Its breath and blood were so toxic that even smelling the creature could prove fatal. What's more, when Hercules chopped off the head of this foul creature, two or three more heads grew back in its place. Hercules and his nephew Iolaus had to burn the neck stumps to stop this happening, and then they were able to defeat the Hydra. In some versions of the story, while battling the Hydra, Hercules also had to contend with being nipped by a crab (see **12. Cancer**).

INTERESTING FACTS

NGC 5694, known as Tombaugh's Globular Cluster, was originally discovered by William Herschel around 1784 but was first defined as a globular cluster by Clyde Tombaugh in 1932. (Tombaugh was, of course, also responsible for the discovery of Pluto two years earlier.) At 12 billion years old, NGC 5694 is one of the oldest known globular clusters in the Milky Way.

43. HYDRUS, THE 'MALE WATER SNAKE'

Pronounced: 'HIGH-druss'
Short: Hyi
Brightest star: β Hydri (RA 0h 25m, Dec. −77°15')

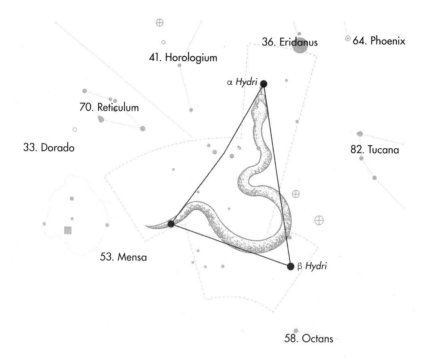

*Hydrus is frequently mistaken for **42. Hydra**, the much larger constellation to the north. Whereas female snake **42. Hydra** is associated with snakes and monsters from Greek mythology, male **43. Hydrus** was based on the sea snakes that Dutch explorers saw on their voyages to the East Indies.*

WHEN AND WHERE TO OBSERVE
From September to November, Hydrus can be seen at latitudes south of +8°. For most observers in the northern hemisphere, this constellation remains permanently below the horizon.

THE BRIGHTEST STARS
Yellow β Hydri is a G-type star with almost the same mass as the Sun but nearly twice the diameter, and it is three times as bright.

White F-type α Hydri was seen by Keyser and de Houtman as the head of the snake constellation, and it's known by that name today in Chinese astronomy: Shé Shǒu or 'snake's head'.

OTHER BODIES

Hydrus has no Messier objects and only a few, faint deep-sky objects visible with large telescopes. One of the most notable is PGC 6240, a very old and large galaxy, also known as the White Rose Galaxy due to its shape. It is approximately 345 million light years away from Earth.

METEOR SHOWERS

There are no significant meteor showers associated with this constellation.

MYTHOLOGIES

There is no mythology surrounding Hydrus. Because the constellation is so far south, it was not visible to the ancient Greeks or Romans. Also, because the stars are relatively dim, there is no record of any other culture including them in their mythologies.

INTERESTING FACTS

This constellation was first depicted on the 1598 celestial globe produced by Petrus Plancius (see **30. Crux**) and Jodocus Hondius, based on observations made earlier that decade by Dutch explorers Pieter Dirkszoon Keyser and Frederick de Houtman (see **22. Chamaeleon**). De Houtman called it De Waterslang, the 'water snake', which was Latinised in 1763 by Nicolas-Louis de Lacaille as Hydrus.

But these different astronomers each had different ideas about the boundaries and shape of the constellation. In fact, it's been said that Hydrus has been redesigned and interpreted more than any other constellation. The current version is shorter than originally conceived; stars that were once thought to be part of it are now assigned to neighbouring constellations.

44. INDUS, THE 'INDIAN'

Pronounced: 'IN-duss'
Short: Ind
Brightest star: α Indi or the Persian (RA 20h 37m, Dec. −47°17')

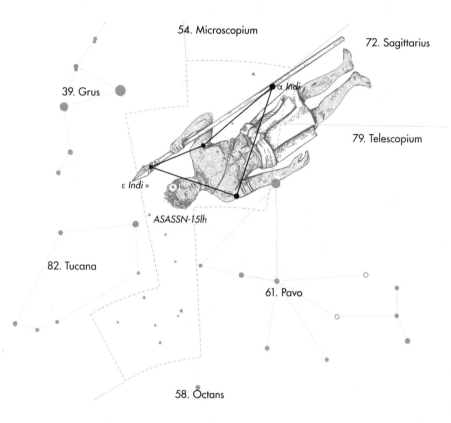

This is another constellation first depicted on the 1598 celestial globe produced by Petrus Plancius (see **30. Crux**) and Jodocus Hondius, based on observations made earlier that decade by Dutch explorers Pieter Dirkszoon Keyser and Frederick de Houtman (see **22. Chamaeleon**). Exploration and mapmaking of this kind aimed to chart safe trade routes to India and the East Indies. This pattern of stars was seen to represent one of the peoples encountered on such voyages: Indus, 'the Indian'.

WHEN AND WHERE TO OBSERVE
Indus is best observed from the southern hemisphere in September, though for much of the year its distinctive triangle can be seen as far north as the equator.

THE BRIGHTEST STARS

Orange α Indi is a giant K-type that has burnt through all of its hydrogen. In China, it's known as Pe Sze, meaning 'Persian' – as in someone from what is now Iran, or the west more generally. This demonstrates how these stars were linked to long-distance trade: Westerners named them after people they met in the East, and Easterners named them after those from the West.

ε Indi is also orange. This is a triple system comprising a K-type star about 75% the size and mass of the Sun, and two T-type brown dwarfs. In orbit around the K-type star is an exoplanet some 3.3 times the mass of Jupiter – the closest Jupiter-like exoplanet we know of.

OTHER BODIES

One object in Indus can only be observed with telescopes but is well worth our attention. The All Sky Automated Survey for SuperNovae (ASAS-SN) is a project in which robotic telescopes around the world automatically scan the sky in search of bright supernovae – exploded stars. In 2015, two ASAS-SN telescopes in Chile detected an object in Indus that was labelled ASASSN-15lh – a boring name for something quite extraordinary.

Although relatively faint when seen from Earth, ASASSN-15lh is thought to be the brightest supernova-like object ever observed. It only appears faint because it's some 3.8 billion light years away. Calculations suggest that, at its peak, ASASSN-15lh was 570 billion times as bright as the Sun and 20 times brighter than the combined light of the entire Milky Way!

That's so extraordinarily bright that scientists aren't sure what ASASSN-15lh actually is. Some suggest it's an especially powerful supernova – a 'hypernova'. Others think it might be a star that was violently torn apart having got too close to a supermassive black hole. There are other theories, all of them weird and wondrous, that we use to try to make sense of this remarkable sight.

METEOR SHOWERS

There are no significant meteor showers associated with this constellation.

MYTHOLOGIES

As with many of these southern constellations, there are no known mythologies associated with Indus.

INTERESTING FACTS

Indus was originally portrayed as a naked male figure with arrows in both hands. Later depictions show Indus with a spear. In the star catalogue *Prodromus Astronomiae* published in 1690 by Elisabeth Hevelius (see **45. Lacerta**), Indus is depicted as a woman.

45. LACERTA, THE 'LIZARD'

Pronounced: 'lah-SER-tah'
Short: Lac
Brightest star: α Lacertae (RA 22h 31m, Dec. +50°16')

A small and faint constellation, this is one of the star groups first defined by the astronomers Elisabeth and Johannes Hevelius. When they named it, they had in mind a specific species of lizard, Stellagama stellio, which has a distinctive pattern on its back thought to look like a line of stars.

WHEN AND WHERE TO OBSERVE

This small, faint constellation sits between the larger and more distinctive **1. Andromeda**, **18. Cassiopeia** and **31. Cygnus**, and the northern part lies on the Milky Way. It's most prominent in the northern hemisphere in October. This constellation is visible between latitudes +90 and −40°.

THE BRIGHTEST STARS

White A-type α Lacertae is roughly twice the mass and diameter of the Sun, and 28 times as bright. Yellow G-type β Lacertae is thought to be a binary system, though it's hard to discern because if there are two components, they are relatively close together and also partly obscured from us by interstellar dust.

OTHER BODIES

There are no Messier objects within this constellation. NGC 7243 is one of the few items of interest that can be seen here. It is an open cluster of stars located approximately 2,800 light years from Earth. This young cluster, thought to be around 100 million years old, is mostly made up of white and blue stars. Unfortunately, as it is quite faint, it can only be seen with a large telescope.

METEOR SHOWERS

There are no significant meteor showers associated with this constellation.

MYTHOLOGIES

As a modern constellation, there are no mythologies associated with Lacerta. However, its brightest stars form a 'W' shape similar to that of mythological queen **18. Cassiopeia**, and it is therefore sometimes referred to as Little Cassiopeia.

INTERESTING FACTS

In 1690, Elisabeth Hevelius published *Prodromus Astronomiae*, a catalogue of 1,564 stars, alphabetised by constellation, which she and her late husband Johannes had spent decades carefully observing from a special platform built on top of their house in Poland.

By this time, astronomy had been radically transformed by a new invention, the telescope, which we think was first used to observe stars in 1609 or 1610 (see **60. Orion**). We know that Elisabeth and Johannes owned telescopes with which they studied the Moon and the planets of the Solar System. Yet for the stars, they worked only by naked eye, using tools such as a quadrant and a sextant (see **77. Sextans**) to take very accurate measurements.

Their catalogue introduced new constellations, filling gaps between those that had been known since the time of Ptolemy. The new constellations are all made up of relatively faint stars, which is probably why they hadn't been recognised as patterns before this. Some think that in making such faint stars into constellations, Johannes was boasting about the quality of his eyesight!

Another six of our 88 constellations were first named in the Hevelius catalogue: **13. Canes Venatici**, **47. Leo Minor**, **51. Lynx**, **75. Scutum**, **77. Sextans** and **88. Vulpecula**.

46. LEO, THE 'LION'

Pronounced: 'LEE-oh'
Short: Leo
Brightest star: α Leonis or Regulus (RA 10h 8m, Dec. +11°58')

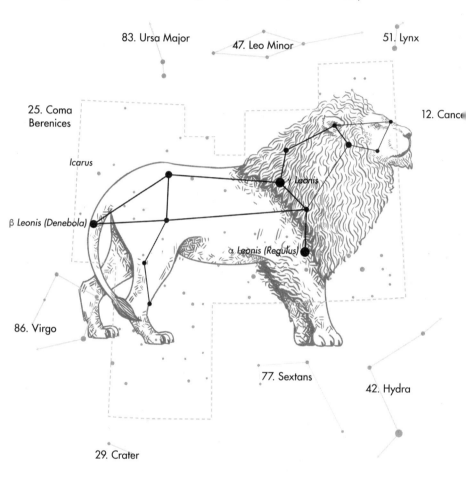

*This is a well-known constellation due to its appearance on the zodiac, and Leo is among the longest-recognised constellations. There's some evidence that it dates back to ancient Mesopotamia in 4000 BCE. Bulls and lions certainly appear prominently in the pottery and carvings of that culture from about 3200 BCE, and sometimes apparently in the sky. There's some debate about whether these artworks depict bull- and lion-shaped constellations (see also **78. Taurus**).*

WHEN AND WHERE TO OBSERVE

This constellation can be easily found using the much more readily apparent Plough or Big Dipper in **83. Ursa Major**: the two stars on the outside of the 'ladle' point towards a pattern of six stars resembling a sickle or inverted question mark. These are the head, neck and front part of the lion. Leo is most prominent in April.

THE BRIGHTEST STARS

Blue-white α Leonis is known as Regulus, or 'prince'. It's actually a system of four stars consisting of two pairs. A blue-white B-type shares a close, 40-day orbit with what we think is a white dwarf with 0.8 times the mass of the Sun. These then share an orbit with one K-type and one M-type star, at a distance of about 5,000 times that between Earth and the Sun. It's estimated that the two pairs of stars take millions of years to complete an orbit of each other!

β Leonis is a blue-white A-type star known as Denebola, from an Arabic phrase for 'lion's tail'.

γ Leonis is a distinctly yellow star that, through a telescope, can be seen to be a binary system, its primary component being a K-type star. Its name Algieba derives from an Arabic phrase meaning 'forehead', though it's actually positioned in the lion's mane.

OTHER BODIES

In 2018, astronomers identified MACS J1149 Lensed Star 1, better known as Icarus, which was at the time the most distant star ever detected, some 14 billion light years from Earth. In August 2022, that record was broken by the discovery of Earendel in **21. Cetus**. We're able to see Icarus at all because its light is magnified by the huge gravitational force of a cluster of galaxies between us – an effect known as 'gravitational lensing'. Icarus is a blue, B-type supergiant, but because of its extreme distance it appears to us as pink!

METEOR SHOWERS

The Leonids meteor shower seems to radiate from near γ Leonis. At its peak around 15 November each year, some ten meteors an hour can be observed, though in 1966 there were reports of 40 meteors a second!

MYTHOLOGIES

The text of the MUL.APIN, inscribed on two clay tablets around 1000 BCE, includes the constellation UR.GU.LA, translated as either 'lion' or 'big dog'. The Babylonians knew its brightest star, α Leonis, as Sharru, meaning 'king', and since the lion has long been considered the king of the animals, that's surely the animal they saw in these stars.

The Babylonian lion seems to have been inherited by cultures across the Middle East and India, with local names for the constellation all meaning 'lion'. To the ancient Greeks, it was the Nemean Lion whose claws could cut through any armour and whose fur repelled all mortal weapons. The hero **40. Hercules** fought and killed the monstrous lion and was often depicted wearing its skin as a cloak that retained this protective ability.

INTERESTING FACTS

The distinctive yellow colour of γ Leonis seems to be the source of its Chinese name, the Twelfth Star of Xuanyuan – the so-called 'Yellow Emperor' who reigned for 100 years until about 2598 BCE.

47. LEO MINOR, THE 'LESSER LION'

Pronounced: 'LEE-oh MY-ner'
Short: LMi
Brightest star: 46 Leonis Minoris or Praecipua (RA 10h 53m, Dec. +34°12')

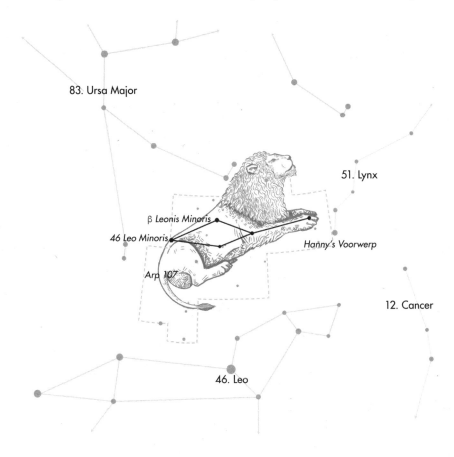

This small, faint constellation was first identified by Johannes and Elisabeth Hevelius and included in the book Elisabeth published in 1690, three years after the death of her husband (see **45. Lacerta**). Leo Minor was clearly created in reference to the larger, older **46. Leo** – one of a number of 'pairs' among our 88 constellations.

WHEN AND WHERE TO OBSERVE
The smaller lion can be found between the more prominent and more easily recognised **46. Leo** and **83. Ursa Major**. As with **46. Leo**, it is completely visible at latitudes north of −48° from January to March.

THE BRIGHTEST STARS

Unusually, there is no α Leonis Minoris. β Leonis Minoris is the only star in the constellation to have a Bayer designation at all. This is a binary star, its primary component a G-type red giant.

However, the brightest star in the constellation is 46 Leonis Minoris. It's likely this was intended to be designated α Leonis Minoris in a star catalogue of 1801 that ascribed such definitions – but for some reason it was left out. This is a K-type red giant with 1.7 times the mass and eight times the diameter of the Sun. It's also known as Praecipua, Latin for 'chief [star in Leo Minor]'.

OTHER BODIES

Only visible with a good telescope, Arp 107 is a group of colliding, merging galaxies some 450 million light years from Earth. Another unusual object in this constellation is Hanny's Voorwerp, discovered in 2007 by Dutch schoolteacher Hanny van Arkel. It's the size of our own Milky Way galaxy, but we think it's the echo of light from a now inactive quasar (itself an extremely bright galactic nucleus powered by a supermassive black hole).

METEOR SHOWERS

The Leonis Minorids are a weak meteor shower seen from 19 to 27 October. At their peak on 23 October, they offer an average of just two meteors an hour, so it's safe to say that there are no significant meteor showers associated with this constellation.

MYTHOLOGIES

Leo Minor is not associated with any myths. It is a new constellation formed from a dark region of the sky. Ancient astronomers thought the region was undefined, with no discernible patterns. Originally, the stars in this area were thought to be part of the constellation **46. Leo**.

INTERESTING FACTS

In 1870, in an attempt to shorten constellation names to make them easier to manage on star charts, English astronomer Richard A. Proctor renamed this constellation Leaena, or 'the Lioness', but the name did not catch on.

48. LEPUS, THE 'HARE'

Pronounced: 'LEEP-uss'
Short: Lep
Brightest star: α Leporis or Arneb (RA 5h 32m, Dec. −17°49')

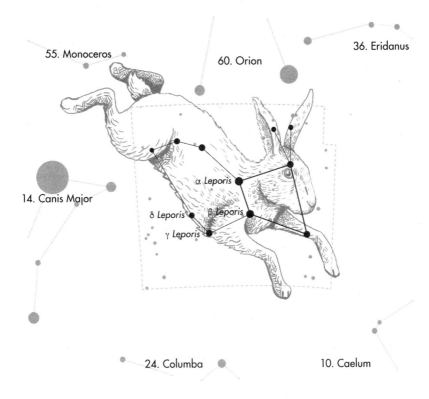

Lepus, meaning 'hare' in Latin, was one of the 48 constellations included in Greek-Egyptian astronomer Claudius Ptolemy's Almagest, *written in about 150 CE.*

WHEN AND WHERE TO OBSERVE

Lepus is located in the skies of the southern hemisphere, but it is visible in the northern hemisphere during winter. It can be seen at latitudes ranging from +63 to −90°. The constellation lies in the northern sky, just under the feet of **60. Orion.**

THE BRIGHTEST STARS

Yellow-white α Leporis is also known as Arneb, from the Arabic for 'hare'. It's an F-type star some 14 times the mass and 75 times the diameter of the Sun. We think it's a supergiant at the end of its life, having burnt through its hydrogen fuel.

β Leporis is a yellow G-type that may be a binary star – we're not sure if its companion is actually very close or just appears that way when seen from Earth. It's known as Nihal, thought to derive from an Arabic phrase meaning 'camels beginning to quench their thirst' – suggesting a completely different interpretation of these stars that has nothing to do with hares!

OTHER BODIES

M79, discovered by Pierre Méchain in 1780, is the only Messier object in Lepus. It is a globular cluster 42,000 light years away from Earth, and one of the few globular clusters visible in the northern hemisphere during winter. It is frequently described as having the shape of a starfish.

IC 418 is a planetary nebula, often called the Spirograph Nebula after the classic drawing toy because of its intricate pattern.

METEOR SHOWERS

There are no significant meteor showers associated with this constellation.

MYTHOLOGIES

The ancient Egyptians saw in both this pattern of stars and neighbouring **60. Orion** a figure called Sah, father of the gods. It was thought that when pharaohs (the kings of ancient Egypt) died, they joined Sah in this part of the sky.

Accounts of Sah – and these stars – were also sometimes mixed up with accounts of Osiris, the Egyptian god of fertility and the dead. The constellation of Orion seems to owe something to Osiris, and the pattern of stars we now call Lepus was thought to be a boat or chair that carried him.

The fact that Osiris was god of fertility may also have had something to do with Lepus being seen as a hare, as various ancient writers noted that hares produce lots of offspring. To Greek poet Aratus (c. 315–240 BCE) this hare constellation was being chased by **14. Canis Major**. In later depictions, this and **15. Canis Minor** are Orion's hunting dogs, and Lepus is their prey.

INTERESTING FACTS

In Chinese astronomy, the stars α Leporis, β Leporis, γ Leporis and δ Leporis are seen not as a hare but as Cè, a celestial toilet.

49. LIBRA, THE 'SCALES'

Pronounced: 'LEE-bruh'
Short: Lib
Brightest star: β Librae or Zubeneschamali (RA 15h 17m, Dec. −9°22')

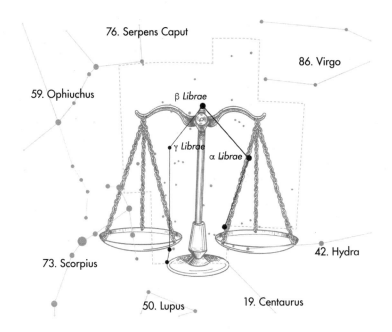

Another well-known constellation due to its appearance in the zodiac, Libra is the only constellation of the zodiac that represents an object rather than an animal or a person from mythology.

WHEN AND WHERE TO OBSERVE
This is a relatively faint constellation in the zodiac, though its brightest stars form a distinctive shape: a triangle with two lines hanging down from two of the corners. Located in the southern hemisphere, it can be seen between latitudes +65 and −90°, and it is visible in the northern hemisphere in winter.

THE BRIGHTEST STARS
The brightest star in Libra, β Librae, is a blue-white B-type – though many observers describe it as looking slightly green. Its name, Zubeneschamali, derives from the Arabic for 'northern claw [of **73. Scorpius**]'.

α Librae is a double star – a blue-white star and a fainter white star can be discerned when seen through binoculars. Its name Zubenelgenubi means 'southern claw'. Orange G-type γ Librae is known as Zubenelhakrabi, meaning 'claws of the scorpion'.

OTHER BODIES

There are no Messier objects in Libra, but there are a few dim deep-sky objects visible with large telescopes. These objects include the barred spiral galaxy NGC 5792, the lenticular galaxy NGC 5890 and the globular star cluster NGC 5897.

METEOR SHOWERS

The May Librids meteor shower occurs between 1 and 9 May each year, with a peak of between two and six meteors per hour visible on 6 May.

MYTHOLOGIES

The earliest known references to Utu, Mesopotamian god of the Sun and of justice, date from about 3500 BCE. He was still being worshipped more than 3,000 years later. The Babylonians knew Utu as Shamash, and saw in this pattern of stars a pair of scales representing his fairness. At the time, the autumn equinox in the northern hemisphere took place while the Sun was in front of these stars, which may also be why the constellation was connected to balance.

By contrast, the ancient Greeks knew the same stars as Chelae, the claws of neighbouring scorpion **73. Scorpius**. Under this name, the constellation appears in the list of 48 constellations given by Greek-Egyptian astronomer Claudius Ptolemy in *Almagest*, written about 150 CE.

INTERESTING FACTS

Around the same time as *Almagest* was written, this constellation was depicted as scales on the earliest known celestial sphere, a two-metre-tall marble sculpture known as the Farnese Atlas. The Romans increasingly referred to the constellation as Libra, Latin for 'scales'.

It used to be claimed that this was due to the influence of Egypt on Rome. Scales figured prominently in ancient Egyptian religion: there are references from at least as far back as 2400 BCE to the gods judging the souls of the dead by weighing a person's heart against a feather. If the heart was lighter than the feather, the person would proceed to the afterlife; if the heart was heavier, their soul was eaten up by the monstrous Ammit.

However, the constellation Libra – as scales – does not appear in Egyptian representations of the night sky until much later. On the Dendera zodiac, carved into the ceiling of a temple in Egypt, we can recognise several of our modern constellations, including **7. Aries**, **16. Capricornus**, **73. Scorpius** and **78. Taurus**, as well as Libra depicted as scales. Various dates have been suggested for the age of this zodiac, but we now think – based on the positions of the planets and eclipses that it also shows – that it was carved about 50 BCE. If this is the case, it's surely a sign that Roman ideas about what the stars represented influenced the Egyptians, and not the other way round!

50. LUPUS, THE 'WOLF'

Pronounced: 'LOOP-us'
Short: Lup
Brightest star: α Lupi or Men (RA 14h 41m, Dec. −47°23')

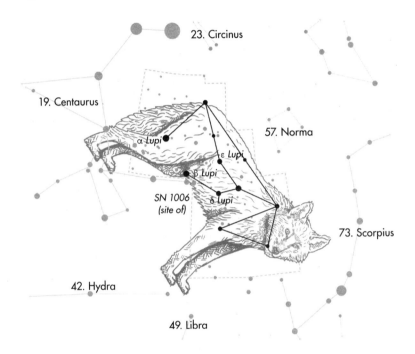

The earliest reference to this wolf-shaped pattern of stars is in **Almagest** by Greek-Egyptian astronomer Claudius Ptolemy, written in about 150 CE. Despite being an ancient constellation, Lupus is not associated with any particular Greek or Roman myths, though it is associated with the family of constellations around the heroic **40. Hercules**.

WHEN AND WHERE TO OBSERVE

Lupus is found in the skies of the southern hemisphere. It is most visible in the northern hemisphere in June and is completely visible at latitudes ranging from +35 to −90°. Lupus is rather in the shadow of its larger, brighter neighbour, **19. Centaurus**, so find that constellation first and you can then locate the more subtle Lupus.

THE BRIGHTEST STARS

α Lupi is a blue giant B-type star some ten times the mass of the Sun and 25,000 times as bright. It's an example of a Cepheid variable (see

20. Cepheus) because its brightness has a regular pulse, varying in apparent magnitude from 2.29 to 2.39 over a seven-hour period.

β Lupi and δ Lupi are also blue giant B-types and Cepheid variables, as are the two brightest components of the multiple star system ε Lupi. These are all thought to be part of a much larger moving group of stars that originated together, the so-called Scorpius–Centaurus Association.

OTHER BODIES

This constellation has no Messier objects, but it does contain a few notable deep-sky objects. NGC 5986 is a globular cluster made up of thousands of stars. It is approximately 33,900 light years away from Earth. The Retina Nebula is a multicoloured planetary nebula; from the side, it appears to be doughnut-shaped. SN 1006 is a supernova remnant created by a bright supernova in the year 1006 CE (see 'Interesting facts', below).

METEOR SHOWERS

There are no significant meteor showers associated with this constellation.

MYTHOLOGIES

The Babylonians saw in this pattern of stars a strange creature with the head and body of a man and the legs and tail of a large carnivore – either a lion, wolf or dog. This may have influenced the ancient Greeks, who named the same stars Therion, after a small creature – such as a wolf or fox – speared by **19. Centaurus** and part of that constellation. Hipparchus of Nicaea (c. 190–c. 120 BCE) described the creature as its own distinct constellation and referred to it as the 'beast'.

Islamic astronomers translated Ptolemy's *Almagest* into Arabic, and Therion became al-Sab, again meaning 'large carnivore', with the constellation depicted variously as either a lion or a wolf. In the 1100s, when Ptolemy's *Almagest* was translated into Latin by Gerard of Cremona, Therion became Lupus, the Latin word for 'wolf' – probably to differentiate it from the lion constellation **46. Leo**.

INTERESTING FACTS

On 30 April 1006, a bright new star was seen in this part of the sky. Egyptian astronomer Ali ibn Ridwan (c. 988–c. 1061) described it as appearing three times as large as Venus and one-quarter as bright as the Moon, lighting up the night sky and remaining visible during the day. The extraordinary sight was recorded by observers all round the world. Some think rock carvings in Arizona made by the Hohokam people are recordings of this phenomenon. According to those who saw it, the 'star' remained visible for months but became progressively fainter. We now think it was an exploding star, known today as supernova SN 1006, whose faint remains have been detected by modern radio telescopes; it is the brightest stellar event ever recorded.

51. LYNX

Pronounced: 'LINKS'
Short: Lyn
Brightest star: α Lyncis (RA 9h 21m, Dec. +34°23')

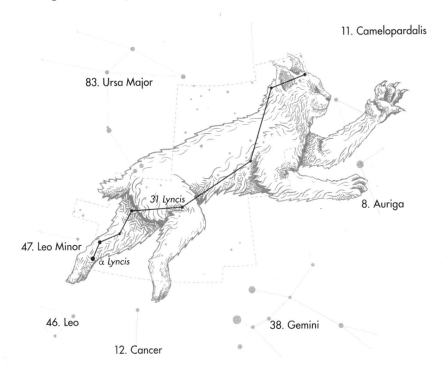

This constellation is located in the northern hemisphere and was intro-
duced in the late seventeenth century by Polish astronomers Elisabeth and
Johannes Hevelius (see 45. Lacerta), who said that the constellation is so
faint it requires the eyesight of a lynx to see it (see 'Interesting facts', below).

WHEN AND WHERE TO OBSERVE
This is a relatively faint constellation whose brightest stars form a jagged
line, but you can find it easily between the more distinct **38. Gemini** and
83. Ursa Major. It's visible in the northern skies throughout much of the
year, but is most prominent on March evenings. It can be seen at latitudes
ranging from +90 to −55°.

THE BRIGHTEST STARS
Orange K-type α Lyncis is the only star in the constellation to have been
given a Bayer designation – a sign of how faint these stars are. Though it

does not appear bright as seen from Earth, it is actually some 673 times as bright as the Sun, and 55 times the diameter: a giant that has swollen up having exhausted its hydrogen fuel.

31 Lyncis is the fourth brightest star in the constellation but the only one with an officially recognised name, Alsciaukat, from an Arabic phrase meaning a 'thorn [in] the outstretched paw'. It's another giant old K-type, this time 53 times the diameter and 782 times the brightness of our Sun.

OTHER BODIES

There are no Messier objects in this constellation, but there are some interesting deep-sky objects.

NGC 2541 is a spiral galaxy with no boundaries. It is 41 million light years away from the Sun and is part of the NGC 2841 group, which includes the Bear Paw Galaxy and other galaxies in the constellations Lynx and **83. Ursa Major**. The Bear Paw Galaxy, also known as NGC 2537, is a blue compact dwarf galaxy – a small galaxy with large clusters of hot young stars that give it a blue appearance.

NGC 2770 is a spiral galaxy that sits 88 million light years away from us. Because four supernova events have been observed in NGC 2770 in recent years, it has been dubbed the 'Supernova Factory'.

METEOR SHOWERS

The September Lyncids are a minor meteor shower that occur around 6 September. Although unimpressive now, records show that they were more prominent in the past. Chinese observers in 1037 and 1063 CE made note of them, as did Korean astronomers in 1560.

MYTHOLOGIES

Johannes and Elisabeth Hevelius created this constellation to bridge the gap between nearby **8. Auriga** and **83. Ursa Major**. As such, there is not much mythology behind the constellation. The Greek-Egyptian astronomer Claudius Ptolemy documented some of the stars in Lynx in the second century, but only as 'unformed' stars near **83. Ursa Major**, not as part of any constellation.

However, there is a weak mythological connection between this constellation and some mythical characters. Lynceus, who sailed with Jason and the Argonauts, is a mythological figure with whom Lynx may be associated. Lynceus was said to have the best eyesight of all men and could even see things from beneath the ground.

INTERESTING FACTS

In about 1612, Petrus Plancius suggested a new constellation reaching from **13. Canes Venatici** to **11. Camelopardalis**. He named it Jordanus after the river Jordan that flows from the Sea of Galilee to the Dead Sea and features several times in the Bible. However, unlike other constellations created by Plancius (see **30. Crux**), this interpretation did not catch on more widely.

One reason was that these stars are rather faint. In their star catalogue published around 1690, Polish astronomers Johannes and Elisabeth Hevelius (see **45. Lacerta**) suggested that one needed to be 'lynx-eyed' to see them, referring to wild cats that are thought to have good eyesight (the name 'lynx' derives from a word meaning 'bright', in reference to the reflective quality of the cats' eyes). In fact, the Heveliuses referred to these stars as both 'lynx' and 'tigris' (Latin for 'tiger'), both of which are known to be difficult to spot in the wild. It was the former that stuck with other astronomers.

52. LYRA, THE 'LYRE'

Pronounced: 'LYE-ruh'
Short: Lyr
Brightest star: α Lyrae or Vega (RA 18h 36m, Dec. +38°47')

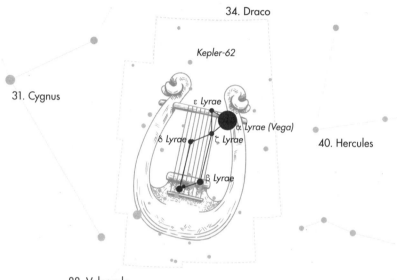

In his **Almagest** of about 150 CE, Greek-Egyptian astronomer Claudius Ptolemy knew both this constellation and its brightest star as the 'Lyre'. The lyre is a stringed musical instrument, the earliest known examples of which are found in Ur in modern Iraq, dating to about 2500 BCE. The earliest known Greek example appears in a picture: a fresco painted on the side of a sarcophagus found at Hagia Triada in Crete dating from about 1400 BCE. The Greeks took the lyre to heart, using it for storytelling and poetry accompanied by music. We still use the word 'lyrics' to describe the words in songs.

WHEN AND WHERE TO OBSERVE
This is a small constellation, but it contains one of the brightest stars in the sky, which makes it relatively easy to locate, not least because of the distinctive Northern Cross shape in its neighbour, **31. Cygnus**. Lyra is visible in northern skies from June to October but is at its most prominent during August. The constellation can be seen between latitudes +90 and −40°.

THE BRIGHTEST STARS
α Lyrae is known as Vega, from an Arabic phrase meaning the 'swooping eagle or vulture' (the same word in Arabic can be used for either bird),

suggesting that this pattern of stars is a partner to the eagle in **5. Aquila**. Several later depictions of this constellation blend the two ideas, with a lyre being carried by such a bird.

Vega is the fifth brightest star in the night sky, and the second brightest in the northern hemisphere (after Arcturus in **9. Boötes**). It's a blue A-type, a little more than twice the mass and diameter of the Sun and 40 times as bright.

β Lyrae is a multiple star system; its two major components are both blue B-type stars. It's known as Sheliak, from the Arabic for 'shell', referring to the legend that the first lyre was made from a turtle shell. The two components are so close that one star is drawing material from the other, creating an accretion disc around the star that, viewed from Earth, causes regular dips in brightness every 13 days.

ε Lyrae is another multiple system and ζ Lyrae is a binary. δ Lyrae is a double star made up of one binary star system and a red-coloured M-type that only appears close to the binary star from our perspective.

OTHER BODIES

Lyra is home to two Messier objects. M57, the Ring Nebula, is the most well known: a colourful planetary nebula formed when a star expelled its outer layer of gas. M56 is a dense globular cluster containing thousands of stars. It is 84 light years across and 32,000 light years away from Earth.

There are a few other deep-sky objects in this constellation, but they are extremely faint and require a large telescope to see them. The most notable is NGC 6745, an irregular galaxy. It's thought this was once a regular spiral, but it was deformed after colliding with another galaxy.

METEOR SHOWERS

The Lyrids meteor shower occurs between 16 and 25 April every year and usually peaks around the night of 22 April. The Lyrids are associated with Comet Thatcher, which orbits the Sun every 415 years. With 15–20 meteors per hour, it is well worth looking out for. Some of these meteors are brighter than usual and are known as Lyrid fireballs.

MYTHOLOGIES

According to Greek mythology, the lyre was invented by the god Hermes, who made the strings from cow gut and the soundbox from a turtle shell. This first lyre was passed on to Orpheus, a musician so skilled that he could charm the rivers and even match the singing of the monstrous sirens, which entranced all those who heard it.

We don't know much about ancient tunes and compositions, but the likelihood is that Orpheus played sad, plaintive music to match the tragic events of his life. When his wife Eurydice suddenly died, Orpheus followed her to the underworld and was allowed to lead her back to daylight – on the condition that he didn't look back at her as they made their journey.

Orpheus played his lyre to lead Eurydice, but she made no sound as she followed him. He began to think that he had been tricked and she wasn't there at all. Finally, he couldn't resist any longer and turned to look – he got his last glimpse of Eurydice as she was swept back to the underworld for good. The grieving Orpheus then wandered the land, playing his sad music. When he died and was reunited with Eurydice, the gods placed his lyre in the night sky.

The constellation Western astronomers know as Lyra forms the base of what the Boorong people called Neilloan, a malleefowl whose outline in the night sky heralded the bird's egg-laying period, as it appears in the southern hemisphere between March and October, coinciding with mound-building season.

INTERESTING FACTS

The Kepler Space Telescope focused on this constellation, **31. Cygnus** and **34. Draco** in its search for planets in orbit round stars other than the Sun. Among its many discoveries, the K-type star Kepler-62 in Lyra is thought to have five planets in orbit around it, including two within the habitable zone, where the temperature and pressure allow for liquid water – and the potential for life.

53. MENSA, THE 'TABLE MOUNTAIN'

Pronounced: 'MEN-suh'
Short: Men
Brightest star: α Mensae (RA 6h 10m, Dec. −74°45')

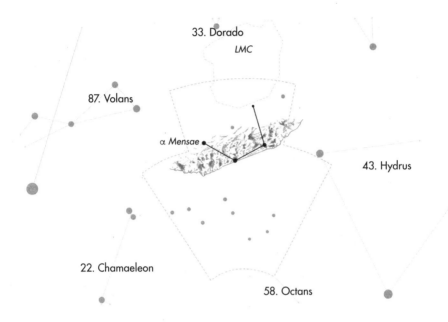

Mensa is a small constellation of dim stars and the only constellation named after a real place on Earth: the huge flat-topped mountain that overlooks Cape Town in South Africa, with a main plateau that stretches some 3km.

WHEN AND WHERE TO OBSERVE

The second most southerly constellation (after **58. Octans**), Mensa cannot be seen north of +4° latitude. Even south of that, it's such a faint constellation that it's very difficult to see except in dark, unpolluted skies. Obviously the ideal place to view this constellation is from the mountain that gives it its name. Plan your trip for January, when the constellation is most prominent in the evening sky.

THE BRIGHTEST STARS

This is the only constellation with no stars brighter than an apparent magnitude of 5. Lacaille listed 11 stars here; the brightest of them – α Mensae – is barely visible to the naked eye, even in dark skies. It's a yellow G-type star, a little smaller, less massive and less bright than the Sun.

OTHER BODIES

This constellation does not contain any Messier objects, but as with its neighbour **33. Dorado**, it contains part of the Large Magellanic Cloud, giving the constellation the appearance of being capped by a white cloud, similar to the clouds that cap Table Mountain.

Mensa also contains an open star cluster called NGC 1987 and a spiral galaxy called IC 2051. Both objects are extremely dim and require a large telescope to see.

METEOR SHOWERS

There are no significant meteor showers associated with this constellation.

MYTHOLOGIES

As with many of the constellations conceived in the modern era, there are no mythologies associated with this constellation.

INTERESTING FACTS

Mensa was devised by French astronomer Nicolas-Louis de Lacaille, who built an observatory in Table Bay in the shadow of the mountain, where he spent two years from 1751 to 1752 observing the stars of the southern sky. This work led to the creation of 14 new constellations (see **2. Antlia**). His planisphere of 1756 included *Montagne de la Table*, created from dim stars to fill a void in the map and named after the place where he'd worked.

A star catalogue published in 1763, the year after Lacaille's death, translated this constellation into Latin as *Mons Mensae*. The English astronomer John Herschel then proposed shortening the name to Mensa, which Francis Baily adopted in his British Association Catalogue in 1845.

People have lived in this area for a very long time: skeletons found at nearby Fish Hoek date from about 5000 BCE, and other evidence may date from much earlier. In 1503, Portuguese sailor António de Saldanha became the first known European to land at what is now Cape Town, and he and his crew climbed the mountain, which they named 'the table of the cape'.

54. MICROSCOPIUM, THE 'MICROSCOPE'

Pronounced: 'my-kroh-SKOPE-ee-um'
Short: Mic
Brightest star: γ Microscopii (RA 21h 1m, Dec. −32°15')

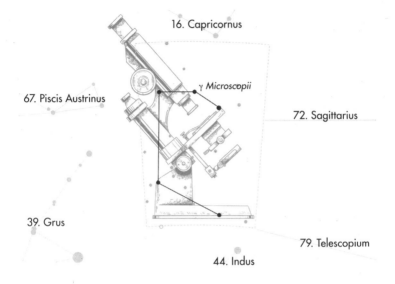

This is a small, faint constellation in the southern sky, barely discernible north of the tropics. To observe this constellation, one should start by spotting its more prominent neighbours, such as **16. Capricornus** and **72. Sagittarius**.

WHEN AND WHERE TO OBSERVE
Microscopium is visible at latitudes south of +45° between July and September and is most visible on September evenings.

THE BRIGHTEST STARS
The brightest star, γ Microscopii, is the eyepiece of the microscope. It's a yellow G-type, 2.5 times the mass and ten times the diameter of the Sun, and 64 times as bright. It's now some 223 light years from Earth, but we think that 3.9 million years ago it passed within 1.1–3.5 light years of the Sun – close enough for its gravitational force to disturb the outer edges of our Solar System.

OTHER BODIES
There are no Messier objects in this constellation, and there are only a few deep-sky objects, which are extremely dim so they can only be seen with

very large telescopes. The most notable of these is Arp-Madore 2026-424, a pair of colliding galaxies. This sounds dramatic, but the space between the stars involved is so vast that it is unlikely any of them will actually collide. Instead, as the galaxies continue to pass through one other, gravity will distort their shapes.

METEOR SHOWERS
There are no significant meteor showers associated with this constellation.

MYTHOLOGIES
As a modern constellation that contains very dim stars, no mythology is known to have been linked to these stars.

INTERESTING FACTS
It might seem odd to see a microscope among the stars – after all, a microscope could be considered to be the polar opposite of a telescope, the instrument we use to study the night sky. But the two instruments do in fact have a lot in common, which is why the name fits this particular constellation.

We've known for a very long time that bending light – with a lens or through water – can make objects appear larger than they are, which is useful for studying them better. Pliny the Elder (c. 23–79 CE) claimed that the Roman Emperor Nero (37–68 CE) used an emerald for such a purpose, while Greek-Egyptian astronomer Claudius Ptolemy wrote a now lost book on optics that included commentary on lenses. Persian mathematician Ibn Sahl (c. 940–1000 CE) and Islamic astronomer Ḥasan Ibn al-Haytham (c. 965–c. 1040 CE) developed many of Ptolemy's ideas, and the earliest known eyeglasses date from the 1200s.

A compound microscope produces greater magnification by using at least two lenses: an objective lens positioned near to whatever is being examined, and an ocular lens close to the eye. The earliest known example dates from about 1620, but we're not sure who invented this kind of microscope.

Among the many candidates is Dutch spectacle-maker Hans Lipperhey (c. 1570–1619), who also claimed, in 1608, to have invented the telescope. Italian astronomer Galileo Galilei (1564–1642) is often also credited as the inventor of the microscope. He was using a telescope to examine planets and stars as early as 1609, and he noticed that the same instrument could be used in reverse to examine small objects. However, the first microscope he built more than a decade after this seems to have been an improved version of someone else's. One reason we don't know who invented the microscope is that for several decades it seems to have been a novelty item.

Then, in the 1660s, there was a sudden boom across Europe in using the microscope to study the natural world. English scientist Robert Hooke published a book in 1665, *Micrographia*, with extraordinary, huge, close-up illustrations of things like a flea, the point of a needle and a piece of cork. It became the first scientific bestseller.

In the 1750s, French astronomer Nicolas-Louis de Lacaille created 14 new constellations (see **2. Antlia**) and named many of them after scientific instruments. Microscopium may get its name because these stars are so small and faint.

55. MONOCEROS, THE 'UNICORN'

Pronounced: 'moh-NOH-seh-ross'
Short: Mon
Brightest star: β Monocerotis (RA 6h 28m, Dec. −7°1')

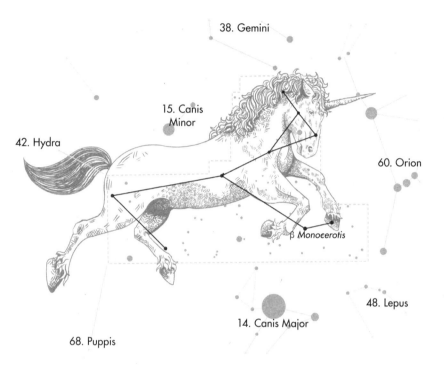

*Despite its lack of bright stars, the Unicorn is easy to locate due to its prox-imity to **60. Orion**. Because the Milky Way passes directly through the centre of the constellation, it is also a rich territory for open star clusters.*

WHEN AND WHERE TO OBSERVE
This faint constellation straddling the celestial equator is visible in the northern hemisphere in winter. It can be seen between latitudes +75 and −90° and is easy to locate between its more prominent neighbours: **14. Canis Major**, **38. Gemini** and **60. Orion**.

THE BRIGHTEST STARS
β Monocerotis is in fact a triple star system; the three blue B-type stars are discernible through a telescope (depending on the focus of the telescope, they can also seem yellow).

OTHER BODIES

M50 is the only Messier object in Monoceros. It is an open cluster of more than 200 stars.

This constellation also contains a large number of fascinating deep-sky objects. One of the most spectacular that I have seen is V838 Monocerotis, which sits about 20,000 light years away from Earth and is one of the most studied bodies in the night sky. That's because, in January 2002, this unassuming star suddenly expanded, causing it to become the brightest star in the entire Milky Way galaxy. Then, just as quickly, it faded away.

We're not sure what caused the V838 Monocerotis outburst, but it was unlike any supernova ever seen before – the first known example of a luminous red nova. As well as ejecting material into space, V838 Monocerotis also flashed brightly. A complex array of interstellar dust already surrounding the star then reflected this radiating light, creating a pattern of rings or 'light echo' some six light years in diameter. The Hubble Space Telescope has produced some extraordinary images of this phenomenon.

METEOR SHOWERS

There are two minor meteor showers associated with this constellation: the December Monocerids and the Alpha Monocerids. The latter occur from 15 to 25 November, with the peak occurring on 21/22 November. Normally, the rate of meteors per hour is low, but it occasionally produces much more intense meteor storms that last less than an hour. Such outbursts were observed in 1925, 1935, 1985 and 1995. The storms of 1925 and 1935 both had hourly rates of over 1,000 meteors!

MYTHOLOGIES

This constellation was created in 1612 or 1613 by Dutch mapmaker Petrus Plancius (see **30. Crux**). By then, the mythological unicorn (a horse- or goat-like creature with a long, straight horn protruding from its forehead) had been commonly depicted in Europe as a woodland creature, so graceful and pure that it could only be captured by people of equal virtue. Unicorn horns were reputed to have purifying powers, making water safe to drink and healing any illness.

The ancient Greeks seem to have thought this was a real animal. The Greek doctor Ctesias of Knidos made the first reference to unicorns in about 400 BCE in a now lost history of India. We don't think Ctesias ever visited India himself, but he reported what he heard while working in Persia (modern-day Iran).

Even in Ctesias' own time, people thought some of these reports rather fanciful, such as his description of one-legged men with feet so big they used them as umbrellas! But he also accurately described elephants a century before they were known in the Greek-speaking world, and his account of unicorns is richly detailed, with later (surviving) texts acknowledging his insightful remarks about unicorn ankle bones.

There's no evidence that unicorns really existed, so where did the idea come from? Cow-like rather than horse-like creatures with single horns are seen in many soapstone seals and other artworks made in the Indus Valley Civilisation dating from around 2000 BCE. Perhaps Ctesias saw or was told about these.

Those cow-like creatures may have been based on real animals. When Venetian merchant Marco Polo (1254–1324) travelled through southeast Asia, he saw 'numerous unicorns'. The creature, he said, had a boar-like head, a thick black horn in the middle of its forehead and liked to wallow in mud. They were, he admitted, nothing like the unicorns of the romantic stories. We now think what he saw was a Sumatran rhinoceros. Perhaps our horse-like unicorn is based on a cow-like missighting of this kind of rhinoceros or some similar animal, and this doubly misunderstood but real creature is what Plancius placed in the night sky.

INTERESTING FACTS
In 2009, two super-Earth exoplanets in a single planetary system were detected in this constellation. The first, COROT-7b, was discovered by the COROT satellite. The second, COROT-7c, was discovered by HARPS ground-based telescopes. Until January 2011, COROT-7b was the smallest exoplanet ever detected, with a diameter just 1.58 times that of Earth.

56. MUSCA, THE 'FLY'

Pronounced: 'MUSS-kah'
Short: Mus
Brightest star: α Muscae (RA 12h 37m, Dec. −69°8')

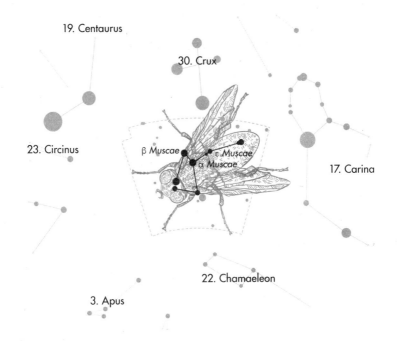

*This small constellation in the southern sky was created in 1597 or 1598 by Dutch mapmaker Petrus Plancius (see **30. Crux**), based on observations made in Madagascar and the Indian Ocean by sailors Pieter Dirkszoon Keyser and Frederick de Houtman (see **22. Chamaeleon**). It was initially unnamed, though in his own work de Houtman referred to it as De Vlieghe, Dutch for 'fly'; the four brightest stars make up the fly's head, body and wings.*

WHEN AND WHERE TO OBSERVE

In the southern hemisphere, this small, faint constellation can be found south of its distinctive neighbour **30. Crux**. It is completely visible in latitudes south of +15° between February and April.

THE BRIGHTEST STARS

α Muscae is a blue-white B-type star, and β Muscae is a binary star composed of two blue-white B-types. All three are part of the Scorpius–Centaurus Association of stars that originated and are moving together.

Several other stars in Musca are part of the same group, as are several of the brightest stars in **50. Lupus**.

One exception is ε Muscae, a red giant M-type star that represents the fly's right wing. At some 300 light years from Earth, it's in the same part of space as the Scorpius–Centaurus Association but has a different origin and is moving at a much greater speed, so over time it will separate from these blue-white stars.

OTHER BODIES

This constellation is devoid of Messier objects, despite its location along the Milky Way.

However, Musca does have notable deep-sky objects. The Spiral Planetary Nebula, NGC 5189, shines with a magnitude of about 8. NGC 4833 is a bright globular cluster with a magnitude of 7.8. The Engraved Hourglass Nebula in Musca is another interesting target to observe – and one that many people have seen before in Hubble Space Telescope images.

Finally, NGC 4372 has a rather foreboding name: the Dark Doodad Nebula. This cloud of gas and dust obscures the stars behind it. It is visible through binoculars and spans the length of six full Moons.

METEOR SHOWERS

There are no significant meteor showers associated with this constellation.

MYTHOLOGIES

There are no known mythologies associated with the stars of this constellation.

INTERESTING FACTS

In 1603, German mapmaker Johann Bayer (1572–1625) included this new constellation in his atlas of the night sky, but he referred to it as Apis, the Latin for 'bee'. This was how it was generally known for more than a century afterwards, but French astronomer Nicolas-Louis de Lacaille (see **2. Antlia**) renamed it to avoid confusion with neighbouring **3. Apus**. Lacaille initially used the French word for 'fly', *la Mouche*, with the Latin version *Musca* used in the star catalogue published after his death in 1763.

Some maps draw dot-to-dot lines between the constellation's stars, which I think make it look less like a fly and more like an axe.

57. NORMA, THE 'CARPENTER'S SQUARE'

Pronounced: 'NOR-muh'
Short: Nor
Brightest star: γ² Normae (RA 16h 19m, Dec. −50°9')

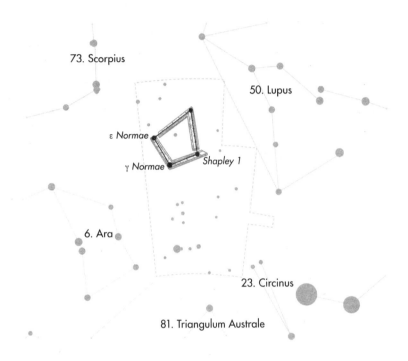

This is one of the **14 constellations** first named in 1756 by Nicolas-Louis de Lacaille (see **2. Antlia**). He initially called it l'Équerre et la Règle, French for 'a set square and a ruler', as used in technical drawing.

WHEN AND WHERE TO OBSERVE

Another faint, small constellation in the southern sky, this is completely visible at latitudes south of +30° from April to June. It is most visible on July evenings, when it can be found due west of the two brightest stars in **19. Centaurus**.

THE BRIGHTEST STARS

In the star catalogue published after Lacaille's death in 1763, Latin versions of the names were applied to his constellations. The set square became Norma – the Latin for 'right angle' – with the 'ruler' part of the name no longer included. Further adjustments to the constellation and the stars it incorporates followed – for example, the stars Lacaille named α Normae and β Normae are now considered to be part of **73. Scorpius**, named N and H Scorpii, respectively.

Such adjustments mean that $γ^2$ Normae is now the brightest star in the constellation: a yellow-orange K-type star. Just discernible beside it is yellow-white F-type supergiant $γ^1$ Normae, but, despite appearances, this is not a binary star – the two components only look close together as viewed from our perspective on Earth. ε Normae is a triple system of blue-white B-type stars.

OTHER BODIES
Visible only by telescope, Shapley 1 is a torus-shaped planetary nebula in orbit round a faint white dwarf star, face on as viewed from Earth – the result being that we see an almost circular cloud with a star at its centre.

METEOR SHOWERS
Visible in the southern hemisphere, the Gamma Normids are a minor meteor shower that usually peak around 15 March.

MYTHOLOGIES
As one of the more recently identified constellations, and due to its small size and faint stars, there are no known mythologies associated with Norma.

INTERESTING FACTS
This is a technical part of the night sky. Beside this constellation, Lacaille created a pair of compasses (**23. Circinus**) and he saw the already created **81. Triangulum Australe** as representing a spirit level.

58. OCTANS, THE 'OCTANT'

Pronounced: 'OCK-tanz'
Short: Oct
Brightest star: ν Octantis (RA 21h 41m, Dec. −77°23')

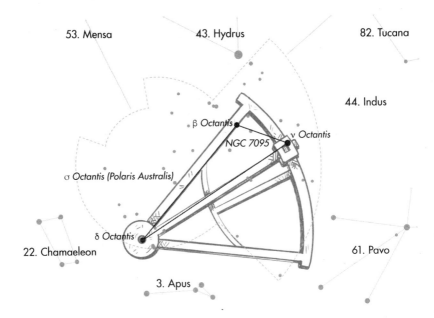

In 1752, French astronomer Nicolas-Louis de Lacaille (see **2. Antlia**) named
this pattern of stars l'Octans de Reflexion, or 'reflecting octant', after a
relatively new device used to accurately measure the height above the horizon of any given celestial body, such as a star.

WHEN AND WHERE TO OBSERVE
The most southerly of constellations – it includes the south celestial pole
– Octans is not visible from the northern hemisphere at all (stood at the
equator, you might just spot it on the horizon), but in the southern hemisphere, this very faint constellation is a constant presence.

THE BRIGHTEST STARS
β Octantis is only barely visible to the naked eye, with an apparent magnitude of 4.13. It's a white A-type star, with some evidence suggesting that it
is a binary system.

σ Octantis is the closest star to the celestial south pole (CSP) and is
sometimes known as Polaris Australis or the southern pole star, the south-

ern counterpart of the better-known Polaris in **84. Ursa Minor.** The CSP is an important marker for navigation, but this F-type star is so faint that it's often not seen directly. Instead, navigators use other, brighter stars to fix its position – such as following an imaginary line from **30. Crux.**

OTHER BODIES

This constellation has no Messier objects and only a few notable deep-sky objects. NGC 7098 is a stunning double-barred spiral galaxy located approximately 95 million light years away from Earth. NGC 7095 is a barred spiral galaxy 115 million light years away. These galaxies are both extremely faint and can only be seen with very large telescopes.

METEOR SHOWERS

There are no significant meteor showers associated with this constellation.

MYTHOLOGIES

As one of the more recently identified constellations, and due to its small size and faint stars, there are no known mythologies associated with Octans.

INTERESTING FACTS

As a tool for accurately measuring the height of a star above the horizon, the reflecting octant was obviously of interest to astronomers, but it also proved essential to sailors, who could use such observations – and the time an observation was made – to calculate their own latitude on the surface of Earth. This massively improved navigation, and the octant was in frequent use until it was superseded by an even better device, the sextant (see **77. Sextans**).

When Lacaille's star catalogue was published in 1763, after his death, the names of his constellations were given in Latin rather than French, and this constellation became Octans Hadleianus, the latter word honouring English mathematician John Hadley, who invented the octant in 1730.

59. OPHIUCHUS, THE 'SERPENT BEARER'

Pronounced: 'OFF-ee-YOO-kuss'
Short: Oph
Brightest star: α Ophiuchi or Rasalhague (RA 17h 34m, Dec. +12°33')

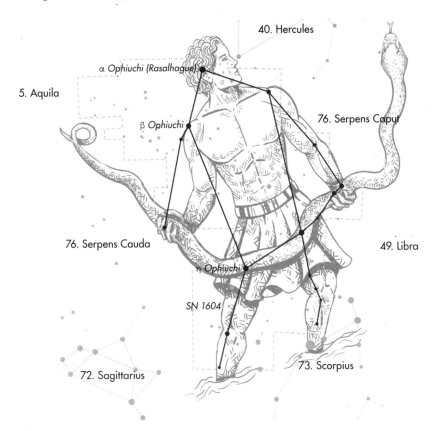

This constellation is sometimes referred to as the 13th sign of the zodiac, and it is one you might not have heard of. In fact, it's one of the oldest-recognised constellations, mentioned by Greek poet Aratus (c. 315–240 BCE), who responded to a description of it in a now lost work by Eratosthenes of Cyrene (276–194 BCE). It is the 11th largest constellation.

WHEN AND WHERE TO OBSERVE
Ophiuchus straddles the celestial equator between two other signs of the zodiac: **72. Sagittarius** and **73. Scorpius**. It is best seen in the northern hemisphere during the summer and in the southern hemisphere during the winter. It is completely visible at latitudes ranging from +80 to −80°.

THE BRIGHTEST STARS

α Ophiuchi is a binary star comprised of one giant white A-type and a smaller K-type, orbiting one another every 8.6 years. It's known as Rasalhague, from an Arabic phrase meaning 'head of the snake charmer' or 'snake collector', suggesting a slightly different interpretation of these stars, yet one still based on a human figure holding a snake.

η Ophiuchi is another binary system, in this case composed of two A-type stars in an unusually close and elliptical orbit. It's known as Sabik, from the Arabic for the 'preceding one' – though we're not sure why.

β Ophiuchi is a giant, distinctly orange K-type star also known as Cebalrai, or the 'shepherd's dog', suggesting the constellation was seen as a shepherd as well as a snake charmer.

OTHER BODIES

The most recent supernova to take place in our galaxy was first seen on 9 October 1604 and was still bright enough to be visible in daylight more than three weeks later. It's known as Kepler's Star after German astronomer Johannes Kepler (1571–1630), who wrote a whole book about it. In the modern system for classifying supernovae, it is SN 1604, and its remnant continues to be much studied by astronomers.

METEOR SHOWERS

The Ophiuchids peak on or around 20 June each year, with an average of 8–20 meteors per hour. As with many showers, on rare occasions many more meteors than this can be seen.

MYTHOLOGIES

According to Greek poet Aratus (c. 315–240 BCE), the constellation represents a man trampling on the eye and chest of monstrous **73. Scorpius**. The snake **76. Serpens** coils around his waist.

Some have suggested that Ophiuchus dates back even earlier than this, the Greeks having drawn inspiration from a Babylonian constellation, Nirah – a god with a human head, body and arms but two serpents for legs.

The Greeks saw in this pattern of stars the god Apollo wrestling with a gigantic snake. Or it was Asclepius, god of medicine, who reputedly learned his skills by watching snakes. Or it was Laocoön, priest of Troy, who warned his people not to accept the Trojan horse presented to them by their enemies, the Greeks; according to legend, Laocoön was later killed by two snakes.

INTERESTING FACTS

In the astrological zodiac, the plane of Earth's orbit round the Sun – known as the ecliptic – is divided into 12 patterns of stars of roughly equal width. As a result, the Sun takes about 30 days to pass in front of each one, more or less matching the length of a month.

But the 13 constellations that overlap the ecliptic – one more than the zodiac because they include Ophiuchus – are not all of the same width. What's more, changes in the rotational axis of Earth mean that the Sun does not pass in front of the same stars as it did thousands of years ago when the signs of the zodiac were first devised, so the constellations of the zodiac in astronomy do not match the astrological signs.

60. ORION, THE 'HUNTER'

Pronounced: 'oh-RYE-un'
Short: Ori
Brightest star: β Orionis or Rigel (RA 5h 14m, Dec. −8°12')

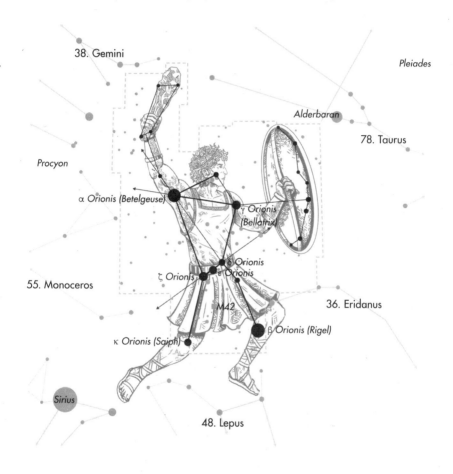

This is one of my favourite constellations. Orion's distinctive belt makes this constellation stand out from the crowd. I like to use it to mark the passing of time from winter to spring, as it can only be observed during winter in the northern hemisphere.

WHEN AND WHERE TO OBSERVE

Across the world, many people notice this distinctive, large, bright pattern of stars. Orion sits above the equator, so it can be seen from both hemispheres at latitudes between +85 and −75°. However, as it sits above the equator, the closer you get to the North or South Pole, the less of Orion you will see – but this is only a problem if you are in the Arctic or Antarctic Circles!

In the southern hemisphere, Orion can be seen around December and January. From the northern hemisphere, Orion also appears in winter as a mighty constellation arcing across the southern part of the sky during the evening hours between November and February.

THE BRIGHTEST STARS

Seven bright stars form the distinctive hourglass pattern. Bright blue-white β Orionis is actually a system of at least four components. Its main star is a B-type star thought to be somewhere between 61,500 and 363,000 times as bright as the Sun. Its name Rigel comes from the Arabic for 'left foot'.

α Orionis is a distinctly large, red star when seen with the naked eye. It is an M-class supergiant more than 760 times the diameter of the Sun. In fact, it's so big that if it were in the Sun's position, it would swallow up Mercury, Venus, Earth and Mars. It's known as Betelgeuse from a phrase meaning 'Orion's hand'.

γ Orionis or Bellatrix ('warrior woman') is a giant blue B-type star. κ Orionis or Saiph (an Arabic sword) is a smaller, fainter B-type.

Orion's belt is composed of triple star system ζ Orionis or Alnitak ('girdle'), blue supergiant ε Orionis or Alnilam ('string of pearls') and multiple star system δ Orionis or Mintaka ('belt').

OTHER BODIES

Sitting below Orion's belt is M42, a stellar nursery called the Orion Nebula. It is big and bright enough to be seen with the naked eye. Closer inspection reveals a richly coloured cloud of gas some 24 light years across, containing clusters of bright, new stars. We have learned a lot about the early life of stars by studying this nebula.

METEOR SHOWERS

The Orionids occur every year between mid-September and mid-November, reaching a peak around 21 October. One of the more impressive showers, it can be seen in both the southern and northern hemispheres. With approximately 15 meteors visible per hour on a moonless night, this is one of the better showers to observe.

The debris trail that forms the basis of this shower comes from the famous Halley's Comet, which can be seen from Earth every 76 years. I was lucky enough to see it as it passed in 1986 and hope to see it on its next visit in 2061.

MYTHOLOGIES

Orion is a strikingly bright constellation often associated with dynamic heroes. To the Chinese, this was the great warrior Shen. To the Babylonians, this figure was the Heavenly Shepherd, or heroic Gilgamesh about to wrestle the bull **78. Taurus**. To the ancient Egyptians it was Sah, father of the gods, and Egyptian kings were thought to join him after death in this part of the sky. In about 750 BCE, the Greek poet Homer referred to the constellation Orion in both the *Iliad* and the *Odyssey*, describing a tall hunter so strong that he wields a club made of solid bronze.

Orion was often linked to other constellations. In some, **14. Canis Major** and **15. Canis Minor** are his hunting dogs, chasing the hare **48. Lepus** or other nearby animal constellations. Or there are the women that Orion pursued, such as the Seven Sisters or Pleiades in **78. Taurus**. In some accounts, his behaviour towards women or his boastfulness about his abilities as a hunter led to a giant scorpion being set upon him; as the constellation **73. Scorpius** rises in the east, Orion 'hides' below the western horizon.

The Orion Nebula, mentioned above, was thought to be the cosmic fire of creation by the Maya of Mesoamerica, and they were quite right to think so: we now know it is a stellar nursery.

INTERESTING FACTS

This may be the longest-recognised constellation of all. A carved fragment of mammoth tusk found in the Ach Valley in Germany dating back to at least 30,000 BCE shows a strident figure with outstretched arms, slender waist and one leg shorter than the other. It's been claimed that this matches the pattern of stars we now know as Orion.

The distinctive stars in Orion can be used as 'pointers' to help us find other constellations. Orion's belt points down to bright Sirius in **14. Canis Major** and up to Aldebaran in **78. Taurus**, and the Pleiades beyond it. Orion's shoulders point the way up to Procyon in **15. Canis Minor**. A line from Rigel and passing through Betelgeuse leads to Castor and Pollux in **38. Gemini**.

But it's worth spending time with Orion itself. Italian astronomer Galileo Galilei constructed his first telescope in 1609, and his sketches of Orion are among the first recorded observations of any stars seen through a telescope. He chose this well-known constellation to demonstrate how much more could be revealed using the new technology. However, he soon gave up on his plan to sketch them all. As he admitted in his book *Starry Messenger* (1610), Orion simply boasts too many wonders.

61. PAVO, THE 'PEACOCK'

Pronounced: 'PAY-voe'
Short: Pav
Brightest star: α Pavonis or Peacock (RA 20h 25m, Dec. −56°44')

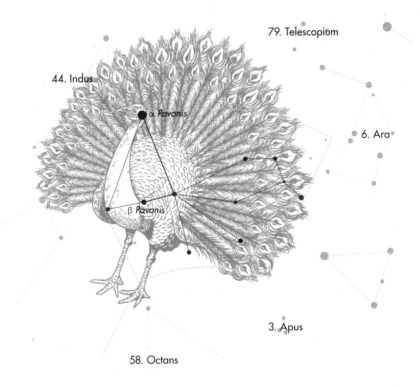

This is one of the Southern Birds, along with **39. Grus, 64. Phoenix** and **82. Tucana**. These birds first appeared on the celestial globe made by Dutch mapmaker Petrus Plancius (see **30. Crux**) and Jodocus Hondius, based on observations made in Madagascar and the Indian Ocean by sailors Pieter Dirkszoon Keyser and Frederick de Houtman (see **22. Chamaeleon**).

WHEN AND WHERE TO OBSERVE

Another small, faint constellation in the southern skies, Pavo is visible at latitudes south of +30° from June to August. It is tricky to spot but is most prominent on August evenings.

THE BRIGHTEST STARS

Blue-white α Pavonis is a binary star, its major component being a B-type thought to be some 2,200 times brighter than the Sun. Evidence suggests its companion is also a B-type, but much fainter and less massive. They orbit one another every 11.75 days.

This star is also unusual for one in a relatively new constellation because it has a name. In the late 1930s, this was one of 57 stars included in the *Air Almanac* produced by the UK's Royal Air Force for use in navigation. All the stars included in the almanac were bright and easily recognisable, and all but two had proper names as well as Bayer designations. The Royal Air Force decided that the remaining two stars should have proper names as well, and W.A. Scott at the Nautical Almanac Office was given the task of naming them. Scott named α Pavonis 'Peacock' and ε Carinae 'Avior'. (The latter, in **17. Carina**, is probably a pun based on 'aviator', the word for an aircraft pilot.)

β Pavonis is a white A-type star thought to have used up its supply of hydrogen and now in the process of becoming a red giant.

OTHER BODIES

There are no Messier objects in this constellation, but there are a few faint galaxies and a globular star cluster.

NGC 6744 is a spiral galaxy that resembles our own Milky Way in shape. The barred spiral galaxy NGC 6782 is 173 million light years away. The Condor Galaxy, also known as NGC 6872, is a barred spiral galaxy with long trailing arms. NGC 6752 is a globular star cluster 13,000 light years away; it is the fourth brightest globular cluster in the sky.

METEOR SHOWERS

There is one weak meteor shower associated with this constellation. The Delta Pavonids occur between 11 March and 16 April, peaking around 30 March. With lowish meteor speed and rates of around five meteors per hour, it is not one of the major showers.

MYTHOLOGIES

The peacock was Hera's sacred bird in Greek mythology; peacocks drew her chariot through the air. Argus, the mythological giant with 100 eyes that could see everything, is associated with the eyes on the peacock's feathers. When Hermes killed Argus, Hera fixed his many eyes on the tail feathers of her peacock.

INTERESTING FACTS

The sailors who first named this constellation seem to have had in mind the green peacock (*Pavo muticus*) that lives in the tropical forests of south-east Asia, which they saw on their travels. It was originally depicted with a more expansive tail – characteristic of peacocks – but this was later cropped back to make space for new constellation **79. Telescopium**.

62. PEGASUS, THE 'WINGED HORSE'

Pronounced: 'PEG-a-suss'
Short: Peg
Brightest star: ε Pegasi or Enif (RA 21h 44m, Dec. +9°52')

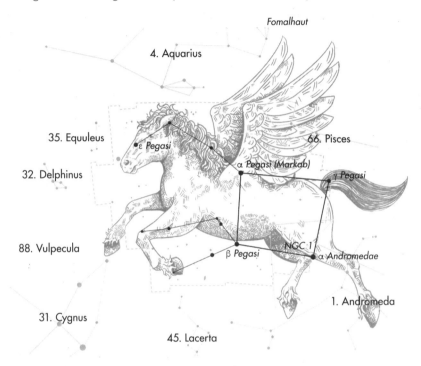

This winged horse is the seventh largest constellation and was included among the 48 constellations listed by Greek-Egyptian astronomer Claudius Ptolemy in his Almagest, *written in about 150* CE.

WHEN AND WHERE TO OBSERVE
This is a large constellation and can be seen between latitudes +90 and −60°. Three of its brightest stars make up the corners of the distinctive asterism known as the Great Square of Pegasus. Its neighbouring constellations are also distinctive, including **1. Andromeda** and **31. Cygnus**, which makes Pegasus relatively easy to find. It's most prominent on October evenings.

THE BRIGHTEST STARS
ε Pegasi is the brightest star in Pegasus, an orange-red K-type supergiant also known as Enif, from the Arabic word for 'nose'. It's thought to be at least 185 times the diameter of the Sun.

β Pegasi is a giant red M-type star known as Scheat, from the Arabic for 'upper arm'. It's another big star, some 95 times the diameter of the Sun. It is also blasting out material, creating an expanding shell of gas currently some 16 times the distance between Earth and the Sun.

OTHER BODIES

A small, round, fuzzy object in Pegasus only visible through a telescope was the first item listed in the 1786 *Catalogue of Nebulae and Clusters of Stars* produced by William Herschel and his sister Caroline, based on years of observations made from their house in New King Street, Bath. The Herschels didn't list it first because they thought it was particularly special: they simply worked in the order that objects appeared in the sky starting from right ascension 0h 0m.

This method continued in subsequent catalogues, including the influential *New General Catalogue* of 1888. As a result, this object is now known as NGC 1. We think NGC 1 is a spiral galaxy some 160,000 light years in diameter. Another even fainter galaxy, NGC 2, can be seen nearby when viewed through a good telescope.

METEOR SHOWERS

The Pegasids occur between 23 July and 3 August each year, peaking around 28 July. This is a weak meteor shower, with a maximum rate of only five high-speed meteors per hour.

MYTHOLOGIES

Greek poet Aratus (c. 315–240 BCE) knew this pattern of stars as Hippos, the 'horse', but it was generally considered to be the mythological Pegasus, a winged horse with magical powers that featured in the story of **63. Perseus** rescuing **1. Andromeda** from monstrous **21. Cetus**.

We know this was the common view because other writers argued against it. Eratosthenes of Cyrene (276–194 BCE) said it couldn't be Pegasus because the horse seen in the night sky did not have any wings. Eratosthenes apparently followed the (now lost) reasoning of playwright Euripides (c. 480–c. 406 BCE) that these stars represented another horse altogether: Melanippe, daughter of Chiron (see **19. Centaurus**).

INTERESTING FACTS

Traditionally, the body of this horse is made up of a square of four bright stars, though one of these is now officially considered to be the brightest star of **1. Andromeda**.

The distinctive Great Square of Pegasus can be used to find other constellations. If the three 'fingers' of the constellation are to the right of the square, the Andromeda Galaxy (in **1. Andromeda**) is above and to the left. A line followed from the right-hand side of the square leads down to Fomalhaut, the brightest star in **67. Piscis Austrinus**.

63. PERSEUS, THE 'HERO'

Pronounced: 'PER-see-us'
Short: Per
Brightest star: α Persei or Mirfak (RA 3h 24m, Dec. +49°51')

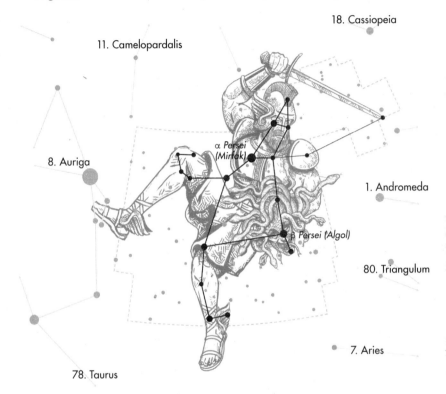

Greek poet Aratus (c. 315–240 BCE) and Eratosthenes of Cyrene (276–194 BCE) both referred to this constellation as Perseus, which is also how it was listed in Almagest by Greek-Egyptian Claudius Ptolemy in about 150 CE. It's a relatively rare example of the different sources all agreeing on the same name and meaning for a given constellation!

WHEN AND WHERE TO OBSERVE

As with **40. Hercules**, Perseus is a famous Greek hero who isn't very bright! The stars that make up the constellation of Perseus are relatively faint, so the best way to find him is to look first for the women that outshine him: **1. Andromeda** (his wife) and **18. Cassiopeia** (his mother-in-law). In the northern hemisphere, the constellation can be seen in late summer and autumn. It is visible from latitudes ranging from +90 to −35°.

THE BRIGHTEST STARS

Yellow-white F-type α Persei is the brightest star in the constellation. Its name, Mirfak, derives from the Arabic for 'elbow'; another name for it, Algenib, is Arabic for 'flank'.

The best-known star in the constellation is the fainter β Persei. This is really a system of three stars – a bright blue-white B-type, an orange K-type and an A-type. As they orbit each other, the B-type and K-type pass in front of and partially eclipse one another as seen from Earth. The result of this is that β Persei regularly pulses over a cycle of 2.87 days. The name Algol comes from the Arabic for 'head of the ghoul', as it was thought to be the severed head of Medusa, still full of monstrous energy even after her death. It's also known as the Demon Star.

OTHER BODIES

Perseus contains two Messier objects. M34 is an open star cluster containing about 400 individual stars. M76, also known as the Little Dumbbell Nebula, is a planetary nebula – an expanding shell of gas ejected from a dying red giant star, which some early astronomers thought looked planet-like. There are also seven stars in Perseus with confirmed exoplanets in orbit around them.

METEOR SHOWERS

The Perseids meteor shower is visible from mid-July each year and peaks around 10–11 August. It's one of the best-known and most reliable meteor showers. The meteors we see are debris left in the wake of Comet Swift–Tuttle burning up in Earth's atmosphere.

MYTHOLOGIES

Perseus was a great hero of Greek mythology. He slew the gorgon Medusa, and from her blood sprang winged horse **62. Pegasus** and its brother Chrysaor. On Pegasus, Perseus rode to the rescue of **1. Andromeda**, in some versions of the story using the gorgon's head to turn the monstrous **21. Cetus** to stone. Andromeda's parents, **18. Cassiopeia** and **20. Cepheus**, also appear nearby in the sky.

INTERESTING FACTS

The Perseus Cluster is a vast neighbourhood of galaxies that share a region of sky with the Perseus constellation. The cluster is located 240 million light years from Earth and stretches 11 light years across. Aside from galaxies, the cluster is mostly made up of extremely hot gases with temperatures reaching tens of millions of degrees. The Perseus Cluster is one of the most massive gravity-bound objects in the known universe, and a region of great scientific interest.

64. PHOENIX

Pronounced: 'FEE-nicks'
Short: Phe
Brightest star: α Phoenicis or Ankaa (RA 0h 26m, Dec. −42°18')

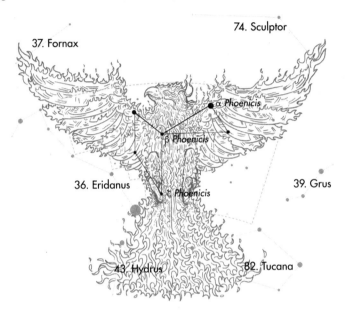

This is one of the four Southern Birds, along with **39. Grus, 61. Pavo** and **82. Tucana**, which first appeared on the 1598 celestial globe made by Dutch mapmaker Petrus Plancius (see **30. Crux**) and Jodocus Hondius. This was based on observations made in Madagascar and the Indian Ocean by sailors Pieter Dirkszoon Keyser and Frederick de Houtman (see **22. Chamaeleon**).

WHEN AND WHERE TO OBSERVE
Another small, faint constellation in the southern skies, from September to November Phoenix is completely visible in latitudes south of +32°, but it appears very low in the sky for anyone north of +40°. Phoenix is most visible on November evenings, ideally in dark skies.

THE BRIGHTEST STARS
α Phoenicis is a binary star comprised of a bright orange K-type and another component of which little is known except that they orbit one another every 10.5 years. Its name, Ankaa, Arabic for 'phoenix', is a modern coinage. (Arab astronomers of the medieval period saw this star as part of a constellation depicting a kind of sailing boat called a dhow.)

β Phoenicis is a binary system of two yellow G-types that orbit one another at a more stately rate of 171 years.

ζ Phoenicis is a multiple star system that, like Algol in **63. Perseus**, has a regular pulse in brightness as seen from Earth because the stars partially eclipse one another. The cycle repeats every 1.6 days. The star is known as Wurren, a 'little fish' that, to the Wardaman culture of northern Australia, provides water for a thirsty porcupine (which they see in nearby Achernar in **36. Eridanus**.)

OTHER BODIES

There are no Messier objects in this constellation, but it does contains many interesting deep-sky objects, such as the irregular dwarf galaxy NGC 625, the lenticular galaxy NGC 37, the group of galaxies known as Robert's Quartet, the HLX-1 luminous X-ray source (which was the first discovered intermediate-mass black hole), the massive Phoenix Cluster and the large, distant cluster of galaxies known as El Gordo ('the Big One').

METEOR SHOWERS

Two meteor showers are associated with this constellation: the December Phoenicids and the July Phoenicids.

MYTHOLOGIES

The other Southern Bird constellations depict real creatures from the East Indies, which Keyser and de Houtman may have seen on their travels there. By contrast, the phoenix is from Greek mythology: a bird reputed to live for hundreds of years before bursting into flames, only to be born again as a young phoenix emerging from the ashes.

The earliest known reference to such a phoenix is in the writings of Hesiod from about 750 BCE, in which its extraordinary life cycle is described by the centaur Chiron (see **19. Centaurus**). The Greek historian Herodotus (c. 484–c. 425 BCE) seemed to think it was a real but rare bird from Egypt, as big as an eagle but with distinctive red and yellow feathers like flames. Other accounts described many different bright colours.

While there's no corresponding tradition of such birds in ancient Egypt, we find similar creatures in the mythology of many Eastern cultures. In fact, depictions found in China of the sunbird *fenghuang* or *Ho-neaou* are thought to pre-date the Greek phoenix, perhaps by thousands of years.

INTERESTING FACTS

Keyser and de Houtman may well have thought the phoenix was a real bird. The first specimens of birds of paradise – brightly plumed species native to Indonesia and Papua New Guinea – are thought to have been brought back to Europe in the 1500s. Several writers of the time thought they were, or were related to, the phoenix of mythology. If so, the name of this constellation is a meeting of science and legend.

65. PICTOR, THE 'PAINTER'S EASEL'

Pronounced: 'PICK-ter'
Short: Pic
Brightest star: α Pictoris (RA 6h 48m, Dec. −61°56')

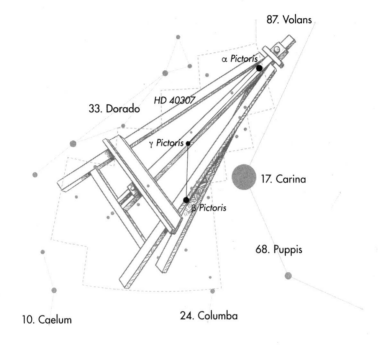

*This small, faint constellation is one of the 14 first named by Nicolas-Louis de Lacaille (see **2. Antlia**). He described it in 1756 as* le Chevalet et la Palette *– 'the easel and palette' of a painter.*

WHEN AND WHERE TO OBSERVE
This small, faint constellation in the southern skies is easy to find beside bright Canopus in neighbouring **17. Carina**. Visible from latitudes south of +26°, it's most prominent in January.

THE BRIGHTEST STARS
The three brightest stars of Pictor are all rather outshone by nearby Canopus in **17. Carina**. α Pictoris is a white A-type. Calculations show that this would be the southern pole star if viewed from the planet Mercury – but we're not planning to view anything from there just yet!

β Pictoris is another white A-type, this one orbited by a disc of dust, ice and rocky debris – it was the first star to be discovered with such a phenomenon, which could be the early stages of a planet being formed.

A planet 12 times the mass of Jupiter (the largest planet in our Solar System) has also been detected in orbit round this star.

γ Pictoris is an orange K-type.

OTHER BODIES
Visible by telescope, HD 40307 is an orange K-type known to have six planets in orbit around it, three discovered in 2008 and three more in 2012. These range from three to ten times the mass of Earth. Five of these planets orbit very close to the star, but the outermost one is, we think, within the habitable zone, where the temperature and pressure are just right for liquid water – and perhaps life – to be present.

METEOR SHOWERS
There are no significant meteor showers associated with this constellation.

MYTHOLOGIES
As one of the more recently identified constellations, and due to its small size and faint stars, there are no known mythologies associated with Pictor.

INTERESTING FACTS
In 1763, a star catalogue based on Lacaille's work renamed the constellations in Latin, so this constellation became Equuleus Pictorius. *Equuleus* literally means 'little horse', as in **35. Equuleus**. Why it's also the word for 'easel' is a bit of a mystery, although a canvas is loaded onto an easel a little like a small horse or donkey might be laden down with things to carry. Whatever the case, since the 1800s the name for the constellation has been shortened to Pictor – which technically means 'painter'.

66. PISCES, THE 'FISHES'

Pronounced: 'PIE-seez'
Short: Psc
Brightest star: η Piscium or Alpherg (RA 1h 31m, Dec. +15°20')

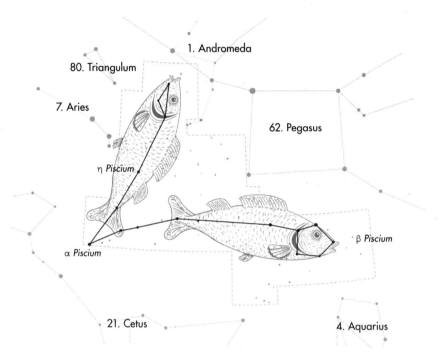

This is a well-known constellation due to being part of the zodiac. In the Babylonian star catalogue MUL.APIN, carved on tablets in about 1000 BCE, this 'V'-shaped pattern of stars is the tail of a huge bird, the swallow. A slightly later text, the DU.NU.NU, described the same 'V' shape as two fish, moving at right angles from one another but with their tails tied together by some kind of ribbon or cord.

WHEN AND WHERE TO OBSERVE
The distinctive 'V' shape of this constellation sits just north of the celestial equator, and the ecliptic passes through it. What's more, the mouth of the 'V' opens towards the distinctive Great Square of **62. Pegasus**, so this is a relatively easy constellation to find. This constellation can be seen at latitudes between +90 and −65°. It is visible in the skies of the northern hemisphere between October and December. It is the 14th largest constellation.

THE BRIGHTEST STARS

η Piscium is a binary star composed of a bright G-type and a fainter B-type. Its name Alpherg is from the Arabic for 'emptying' or 'pouring point'.

α Piscium is a binary star composed of two A-types that take more than 3,000 years to orbit one another. The traditional name Alrescha derives from the Arabic for 'cord'. Blue-white B-type β Piscium is also known as Fumalsamakah, meaning 'mouth of the fish' – just as with Fomalhaut ('mouth of the whale') in **67. Piscis Austrinus.**

OTHER BODIES

Pisces contains one Messier object, M74. Also known as the Phantom Galaxy (NGC 628), this is a stunning face-on spiral galaxy 32 million light years away from Earth. It is very dim, making it challenging for astronomers with amateur equipment to observe it. It is estimated to have 100 billion stars.

Several other faint galaxies can also be found here. NGC 488 is the most notable of these, a spiral galaxy with tightly wrapped spiral arms and dark dust lanes.

METEOR SHOWERS

The Piscids are an annual meteor shower seen throughout September, generally peaking on 20 September. At this point, as many as ten meteors per hour are visible if viewed from a dark location.

MYTHOLOGIES

The ancient Greeks knew this constellation as Ichthyes, the 'fish'. Greek poet Aratus (c. 315–240 BCE) referred to the tails of two fish being knotted together rather than tied with a cord, suggesting that the Greeks inherited this constellation from the Babylonians.

According to Greek mythology, the goddess Aphrodite and her son Eros escaped from monstrous Typhon by leaping into the river Euphrates and turning themselves into fish, tying their tails together so that they would not lose one another as they swam away. (A very similar story was told about the satyr Pan, linked to the constellation **16. Capricornus.**)

One of those who recounted this story was Roman scholar Gaius Julius Hyginus (c. 64 BCE–17 CE), but he also told another version about an egg that rolled into the river Euphrates. Some friendly fish got this egg safely back to shore, where Aphrodite hatched out of it.

Whatever the merits of these conflicting stories, the shared location is worth noting. The Euphrates does not flow through Greece; it originates in what is now Turkey, passes through Syria and Iraq and joins the river Tigris before reaching the Persian Gulf. The ancient city of Babylon was built on its banks, and it's thought the references to the Euphrates in these Greek legends are evidence that they – and the constellation – have Babylonian roots.

INTERESTING FACTS

In German folklore the fish in this constellation are related to the story of Antenteh, a man of limited means who only owned a modest bath and a cabin. He met two magical fish, who offered to grant him any wish. Antenteh initially declined, but, encouraged by his wife, asked the fish for a luxuriously furnished home. That wish was granted, but soon Antenteh's wife wanted more: she wanted to be a queen and to have a palace, and then she asked to be a goddess. At this, the fish became enraged and returned both Antenteh and his wife back to their original status, with just a bath and a cabin once more. The bath is thought to be depicted in the Great Square asterism of **62. Pegasus**.

67. PISCIS AUSTRINUS, THE 'SOUTHERN FISH'

Pronounced: 'PIE-seez oh-STRY-nus'
Short: PsA
Brightest star: α Piscis Austrini or Fomalhaut (RA 22h 57m, Dec. –29°37')

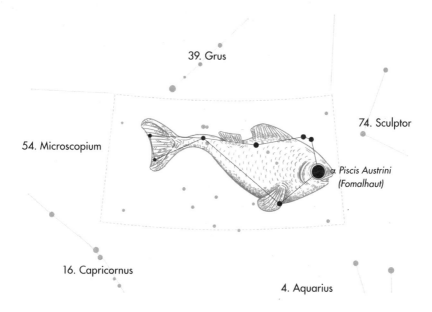

Piscis Austrinus is one of the 48 constellations catalogued in the second century by the Greek-Egyptian astronomer Claudius Ptolemy. It lies in the southern sky.

WHEN AND WHERE TO OBSERVE

This constellation is visible at latitudes south of +53° from July to September. It is most prominent during October.

This is not a large or a bright constellation, but its one bright star, Fomalhaut, can be easily found by following an imaginary line down from one side of the Great Square of **62. Pegasus**.

THE BRIGHTEST STARS

The constellation's only bright star is α Piscis Austrini, or Fomalhaut, from an Arabic phrase meaning 'mouth of the whale'. It is a relatively large constellation, but we think the whale part is simply a mistranslation from Ptolemy.

Several things make Fomalhaut noteworthy. First, this bright A-type star has been used since 1943 as one of the 'anchor points' from which we construct the stellar classification of all stars.

Second, the star is orbited by several different rings of dust and debris. In 2012, it was thought that the Hubble Space Telescope had taken the first ever direct image of an exoplanet here – one larger than Neptune and just inside the outermost ring. In 2015, this planet was named Dagon after a half-man, half-fish god worshipped in ancient Syria and the Middle East. However, the more we've learned about Dagon, the more it seems it might be something stranger than a planet. The current leading theory is that it is a vast, expanding dust cloud.

OTHER BODIES
This constellation contains no Messier objects, but it does have a few interesting deep-sky objects. The most notable of these are a Seyfert spiral galaxy (NGC 7314) and a trio of elliptical galaxies (NGC 7173, NGC 7174 and NGC 7176). They are extremely dim so can only be seen with a large telescope.

METEOR SHOWERS
The Piscis Austrinids are usually active from 15 July to 10 August, reaching their peak around 29 July. This is a medium-density meteor shower, with around five meteors an hour at its peak.

MYTHOLOGIES
The Babylonians knew this pattern of stars as MUL.KU, the 'fish'. As with many Babylonian constellations, this was then passed on to and adapted by the Greeks. In Greek mythology, a Great Fish was depicted as swimming in or drinking the water poured out by **4. Aquarius**. The two fish of the Pisces constellation were said to be the Great Fish's offspring.

That seems to overlap with the ancient Egyptian belief that this was the great fish and its offspring who saved the life of the goddess Isis and were rewarded with eternal life in the night sky. But the Greek historian Ctesias wrote in the fifth century BCE that this fish saved Derceto, a Syrian goddess associated with Aphrodite. As a result, claimed Ctesias, the people of Syria thought fish were sacred and made images of them out of gold.

INTERESTING FACTS
Ptolemy identified this constellation as the 'southern' fish to differentiate it from **66. Pisces**. They were once much more equal in size, but in 1597 or 1598, Dutch mapmaker Petrus Plancius (see **30. Crux**) took some of what had once been the southern fish's tail and included it in his new constellation **39. Grus**.

68. PUPPIS, THE 'STERN'

Pronounced: 'PUP-iss'
Short: Pup
Brightest star: ζ Puppis or Naos (RA 8h 3m, Dec. −40°0')

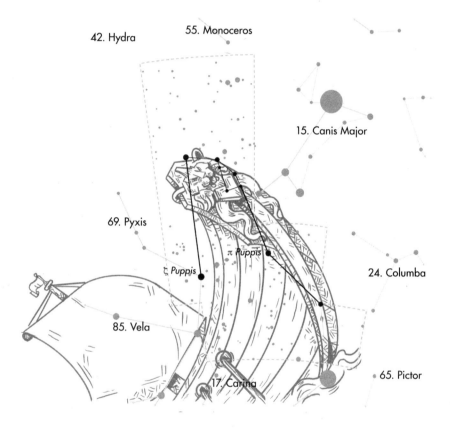

*Puppis is the poop deck or stern of Argo Navis, the enormous boat-shaped constellation recognised by Claudius Ptolemy in Almagest but broken up into more manageable pieces in 1752 by French astronomer Nicolas-Louis de Lacaille (see **17. Carina**, the 'Keel', for more details; **85. Vela** is the boat's sails, and Lacaille then added **69. Pyxis**, the boat's compass).*

WHEN AND WHERE TO OBSERVE

This relatively small, faint constellation in the southern sky borders the more distinctive **14. Canis Major** and its brightest star, Sirius. It's most visible on February evenings.

THE BRIGHTEST STARS

Lacaille chose to keep the Bayer designations applied to Argo Navis, so the brightest star in this constellation is ζ Puppis, or Naos, from the Greek word for 'boat'. This blue supergiant is a relatively rare example of an O-type – one of the hottest, brightest stars – visible to the naked eye. In the visual part of the spectrum, it is more than 10,000 times brighter than the Sun, but that's just what we can see.

With a surface temperature of more than 40,000K, most of the energy this star radiates is in the invisible, ultraviolet part of the spectrum. We actually think it is releasing 800,000 times the energy of the Sun, shedding material at a rate more than 10 million times that of the Sun. This radiates out as a powerful stellar wind moving at approximately 2,500km/s.

π Puppis is a much more typical orange K-type. Having burnt through its hydrogen fuel, it has expanded to some 290 times the diameter of the Sun but remains a relatively cool 4,000K (that's still hot enough to melt carbon!).

OTHER BODIES

There are many open clusters in Puppis because the Milky Way runs through it. M46 and M47 can both be seen in the same binocular field. Under dark skies, M47 can be seen with the naked eye, and its brightest stars are around magnitude 6.

M93 is an open cluster that sits slightly further south. The nearby NGC 2477 is good for viewing with small telescopes, and NGC 2451 is a bright open cluster that contains the star c Puppis.

METEOR SHOWERS

Three meteor showers are associated with this constellation. The Zeta Puppids run from 2 November to 20 December, peaking on 13 November – but with only about three meteors per hour. The Puppid-Velids offer a slightly higher rate of meteors, and can be seen from 2 to 16 December, peaking on 12 December. Finally, the Pi Puppids occur between 15 and 28 April, peaking on 23 April – but they can vary in intensity greatly.

MYTHOLOGIES

See **17. Carina**.

INTERESTING FACTS

Puppis is a large constellation, the 20th of 88 in size, so the original constellation of Argo Navis must have been very big indeed.

69. PYXIS, THE 'MARINER'S COMPASS'

Pronounced: 'PICKS-iss'
Short: Pyx
Brightest star: α Pyxidis (RA 8h 43m, Dec. −33°11')

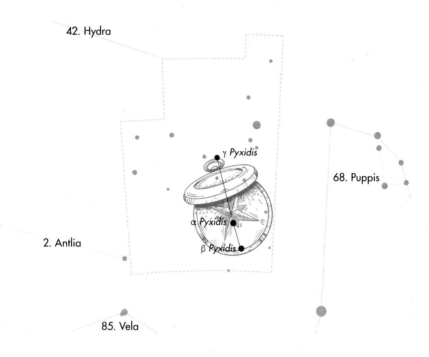

This is one of the 14 constellations first named by French astronomer Nicolas-Louis de Lacaille (see **2. Antlia**). In 1752, he called it Boussole, after a kind of magnetic compass used to find north by sailors. This name was translated into Latin in 1763 as Pixis Nautica – the 'nautical compass' – later abbreviated to Pyxis. Sailors, of course, used compasses like this and measurements of the positions of the visible stars to navigate their way across the oceans. Surely there must have been times when a sailor made use of the marine compass constellation and an actual marine compass at the same time!

WHEN AND WHERE TO OBSERVE
This is a very small, faint constellation in the southern sky and many of its neighbours are small and faint, too. The best way to locate it is first to identify the other parts of the greater boat Argo Navis (see **17. Carina**). Pyxis is most prominent in March.

THE BRIGHTEST STARS

Blue-white α Pyxidis is a giant B-type star with a little more than ten times the mass of the Sun. It is six times the diameter, too, and an extraordinary 10,000 times as bright. In fact, we only see about 70% of its light from Earth, as the rest is obscured by interstellar dust.

Yellow β Pyxidis is a double star. The main component is a bright, giant G-type with an unusually fast spin rate for a star of this kind. One theory to explain this spin is that the star has recently swallowed up a huge planet, such as a so-called hot Jupiter.

Orange γ Pyxidis is a K-type star nearly 22 times the diameter of the Sun but with just 1.6 times the mass.

OTHER BODIES

Pyxis does not have any Messier objects, but it does have a few interesting deep-sky objects. However, these are extremely faint and can only be seen with large telescopes. The planetary nebula NGC 2818 was formed when a dying star ejected its outer layers of gas into space. NGC 2613 is a barred spiral galaxy visible almost edge on from our perspective. The Pyxis globular cluster is made up of thousands of stars; it is one of the most distant globular clusters from the centre of our Milky Way.

METEOR SHOWERS

There are no significant meteor showers associated with this constellation.

MYTHOLOGIES

See **17. Carina**.

INTERESTING FACTS

In the 1800s, some English astronomers proposed that Lacaille's Pyxis should be renamed Malus – the 'mast' of the great boat.

70. RETICULUM, THE 'RETICLE'

Pronounced: 'reh-TICK-yoo-lum'
Short: Ret
Brightest star: α Reticuli (RA 4h 14m, Dec. −62°28')

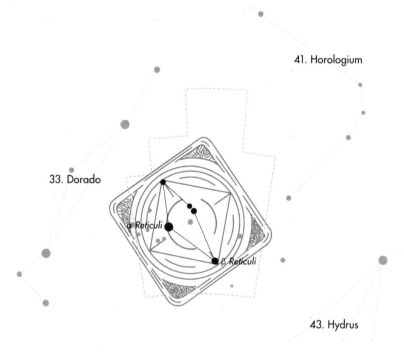

This is the seventh smallest constellation in the night sky and it does not have any bright stars. It depicts a small net of crosshairs on the eyepiece of a telescope used to measure star positions.

WHEN AND WHERE TO OBSERVE
This small, faint constellation in the southern skies is difficult to spot. It's not visible at all above latitudes of +30°, and between there and +23° conditions have to be excellent to pick out such faint stars. It doesn't help that its neighbouring constellations are all faint, as well. But for those up to the challenge, Reticulum is most visible in January. Good luck!

THE BRIGHTEST STARS
Yellow α Reticuli is a G-type star with three times the mass of the Sun. β Reticuli is a binary system composed of a K-type and a fainter companion, thought to be a K-type red dwarf.

OTHER BODIES

There are no Messier objects in this constellation, but there are a couple of notable deep-sky objects. These are extremely dim and can only be seen with large telescopes.

The Topsy Turvy Galaxy (NGC 1313) is a barred spiral galaxy discovered by the Scottish astronomer James Dunlop in 1826. It has a diameter of about 50,000 light years, or about half the size of the Milky Way. Its shape is irregular and its axis of rotation is not in the centre.

NGC 1559 is a barred spiral galaxy about 100 times less massive than our own Milky Way. It contains areas of intense star formation.

METEOR SHOWERS

There are no significant meteor showers associated with this constellation.

MYTHOLOGIES

As one of the more recently identified constellations, and due to its small size and faint stars, there are no known mythologies associated with Reticulum.

INTERESTING FACTS

In 1621, French astronomer Isaac Habrecht II (1589–1633) proposed a new constellation in this part of the sky that he called Rhombus, the technical term known since ancient times for any shape with four sides of the same length, including the diamond and square.

Then, in 1752, another French astronomer, Nicolas-Louis de Lacaille, created a smaller constellation incorporating some of the same stars. It was one of 14 new constellations he created, all but one of them named after technical equipment (see **2. Antlia**). But surely Lacaille acknowledged some debt to Habrecht by calling this one *le Réticule Rhomboide*, after the rhombus-shaped reticle or pattern of lines in the eyepiece of the very telescope he used to observe the stars. Reticulum is the Latin form of the name.

Of course, this means Lacaille was looking through such a reticle to study this reticle-shaped constellation. In doing so, he would have noted the time of his observations to chart the passage of the stars through the sky, and therefore used a clock like the neighbouring **41. Horologium** – a constellation he also created.

71. SAGITTA, THE 'ARROW'

Pronounced: 'sah-JITT-ah'
Short: Sge
Brightest star: γ Sagittae (RA 19h 58m, Dec. +19°29')

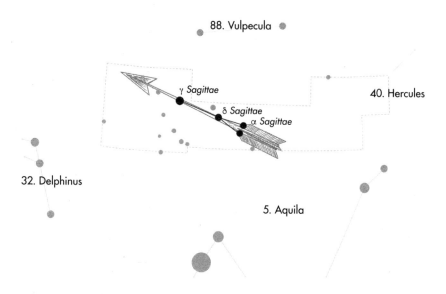

Greek poet Aratus (c. 315–240 BCE) knew this constellation as Oistos, the 'arrow', and that name was used by Greek-Egyptian astronomer Claudius Ptolemy when he included it among the 48 constellations listed in Almagest *in about 150 CE.*

WHEN AND WHERE TO OBSERVE

This is the third smallest constellation, and it sits not very brightly in the northern sky. Yet its arrow shape makes it distinctive, especially if you imagine it has been fired by nearby **40. Hercules**. Sagitta is visible through the summer and is most prominent in August.

THE BRIGHTEST STARS

γ Sagittae is the brightest star in this constellation. It is a red giant about 274 light years away from Earth, and it has 640 times the brightness of the Sun. Next in brightness is δ Sagittae, a binary star composed of a red M-type giant and a blue-white B-type star.

α Sagittae is the third brightest star in the constellation. It's a yellow G-type star with about four times the mass of the Sun, and it is known as Sham, based on the Arabic word for 'arrow'.

OTHER BODIES

Sagitta is home to one Messier object and several notable deep-sky objects. M71 is a globular star cluster composed of at least 20,000 stars. It spans 27 light years and is 13,000 light years away from Earth.

The Necklace Nebula is a huge planetary nebula some 19 trillion kilometres wide, located about 15,000 light years away from Earth. NGC 6886 is also a planetary nebula in this constellation, discovered by English astronomer Ralph Copeland in 1884. It is composed of a hot central star that has around half of the Sun's mass but 2,700 times its luminosity, with a surface temperature of 142,000K.

METEOR SHOWERS

There are no significant meteor showers associated with this constellation.

MYTHOLOGIES

A surviving text of Aratus' book *Phenomena* was translated into Latin by Germanicus Caesar (15 BCE–19 CE), the nephew of Roman Emperor Tiberius. According to Germanicus, the arrow constellation – Sagitta, in Latin – was fired by the god of love Eros at Zeus. This made Zeus fall in love with Ganymede, a shepherd it has been said is represented by **4. Aquarius.** Germanicus said the arrow was then carried into the night sky by an eagle, **5. Aquila.**

Around the time Germanicus wrote this, another Roman author, Gaius Julius Hyginus (c. 64 BCE–17 CE), claimed that this was in fact the arrow with which heroic **40. Hercules** killed the eagle (again, **5. Aquila**) that he saw feasting on the liver of Prometheus. It's true that Sagitta sits between these two constellations, but it's surely not facing in the right direction to have been 'shot' from one to the other.

Eratosthenes of Cyrene (276–194 BCE) knew the same constellation as Toxon, which means 'bow', and claimed it was the bow used by Zeus to kill the Cyclops. Other sources claim that it was Apollo who used it to kill the Cyclops, or that **40. Hercules** shot such an arrow at the Stymphalian birds.

INTERESTING FACTS

Sagitta can be seen from most places on Earth, but not within the Arctic Circle surrounding the North Pole, so here's an odd thought: this is a constellation that polar bears don't see.

72. SAGITTARIUS, THE 'ARCHER'

Pronounced: 'sah-jih-TARE-ee-us'
Short: Sgr
Brightest star: ε Sagittarii or Kaus Australis (RA 18h 24m, Dec. −34°23')

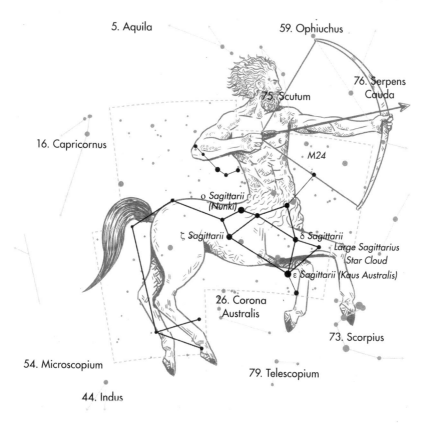

*Sagittarius is typically shown as a centaur pulling back on a bow. This ancient constellation is thought to have been created by the Babylonians. Sagittarius was their god of war, and he stands with his bow aimed at the heart of **73. Scorpius**.*

WHEN AND WHERE TO OBSERVE

The brightest stars of Sagittarius make up an asterism affectionately known as the Teapot, which makes it very easy to spot, especially in relation to other constellations in the zodiac. In dark skies, the Teapot is even billowing 'steam', in the form of the diffuse band of stars that make up the Milky Way! The archer is located in the southern hemisphere. It can be seen in the northern hemisphere in the summer as far north as +55°, and in the southern hemisphere in winter.

THE BRIGHTEST STARS

ε Sagittarii is the brightest star in the constellation and is known as Kaus Australis; Kaus comes from the Arabic for 'bow'. We now know this is a binary system, the main component being a blue B-type with a wide disc of dust orbiting round it. Its companion star is less than one-third the mass of the B-type, and we think it is located within the disc of dust.

σ Sagittarii is the second brightest star in Sagittarius, a B-type also known as Nunki – an Assyrian or Babylonian name, though we're not sure of its meaning. We think this is the oldest name for a star still in use.

ζ Sagittarii is also known as Ascella, a Latin word for 'armpit'. It's really a triple system comprised of two A-type stars and a much fainter companion.

δ Sagittarii is a binary star comprised of a giant K-type and a white dwarf. It's known as Kaus Media, or the 'middle of the bow'.

OTHER BODIES

Seen from Earth, the centre of our galaxy is in Sagittarius. This means the Milky Way is at its most dense in this part of the night sky, and many star clusters, nebulae and other deep-sky objects can be seen through binoculars and telescopes here. The Large Sagittarius Star Cloud and the Small Sagittarius Star Cloud (M24) are a good place to start!

METEOR SHOWERS

The Sagittarids occur between 1 June and 15 July, with medium-level peak activity generally on 19 June.

MYTHOLOGIES

In the Babylonian star catalogue MUL.APIN of about 1000 BCE, this pattern of stars represents a god called Pabilsaĝ, while in other Babylonian texts it is a different god, Nergal. In both cases, the god was seen to be drawing the string of a great bow comprising the brightest of the stars here. The rest of the constellation – the god's body – was less clearly visible, which may explain variations in the way it was depicted. It usually had the head, torso and arms of a human, but there could be a second head (sometimes that of a panther) and the body of a horse or a bull, great wings and/or a sting like that of a scorpion.

The ancient Greeks seem to have inherited something of this conception, simplifying it to a half-man, half-horse centaur. Greek poet Aratus (c. 315–240 BCE) described it as such, as did Claudius Ptolemy in *Almagest* in about 150 CE, who described the figure's billowing cloak (where the Babylonians had seen wings or a sting). Some accounts suggested that this figure is Chiron, wise centaur of Greek mythology, but he is also associated with **19. Centaurus**.

Therefore, this is either another centaur, or not a centaur at all. Eratosthenes of Cyrene (276–194 BCE) argued for the latter, claiming that it only had two legs rather than four, and was thus a half-goat, half-man called a satyr.

Whoever this archer might be, the bow and arrow are pointed directly at **73. Scorpius**, and several explanations were given for this, such as the archer responding to the scorpion attacking either **40. Hercules** or **60. Orion**.

INTERESTING FACTS

The New Horizons space probe, which launched in 2006 and flew by Pluto in 2015, is now heading in the direction of Sagittarius – though it will have exhausted its fuel long before it reaches the nearest of the stars there.

73. SCORPIUS, THE 'SCORPION'

Pronounced: 'SKOR-pee-us'
Short: Sco
Brightest star: α Scorpii or Antares (RA 16h 29m, Dec. −26°25')

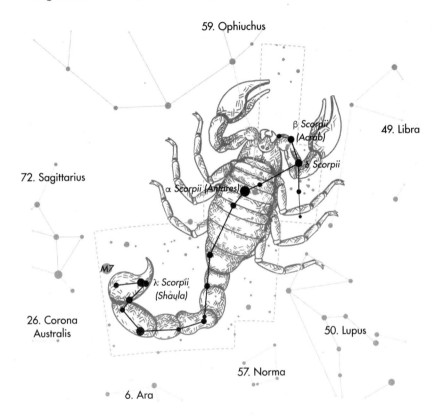

From about 3200 BCE, the Sumerian and Babylonian civilisations in what are now Iraq and Iran produced a great wealth of art and writing. Among the items that survive are carved pottery and seals that include depictions of animals. In some of these, scorpions, lions and bulls appear to be suspended in the sky. If, as we think, these correspond to the modern Scorpius, **46. Leo** and **78. Taurus**, these must be among the oldest-recognised constellations.

WHEN AND WHERE TO OBSERVE

Hook-shaped Scorpius comprises many bright stars and is easily found in relation to other constellations in the zodiac. It can be seen at latitudes ranging from +40 to −90°. Located in the sky's southern hemisphere, it can be seen from the northern hemisphere during the summer, but it is low in the sky and is best seen from the southern hemisphere.

THE BRIGHTEST STARS

The distinctly red-coloured α Scorpii is also known as Antares, from the Greek for 'rival to Ares' – the Greek name for the red planet Mars, which looks similar when seen by the naked eye. In fact, there is little contest between the two. This is a red supergiant M-type star, thought to be about 12 times as massive as the Sun; if it were located where the Sun is, Mercury, Venus, Earth and Mars would all be inside it.

λ Scorpii is the second brightest star in the system (despite its Bayer designation as λ). It's a triple system, two of its components being bright B-types. The name Shaula comes from the Arabic for 'sting', though the word may link back to the Babylonian MUL.APIN, where this star is named Sharur and Shargaz.

β Scorpii – known as Acrab, from the Arabic for 'scorpion' – is a multiple system of at least six stars. The two largest components are B-types some 15 and 10 times as massive as the Sun.

δ Scorpii is a binary star system and possibly a triple one. The main component being a B-type star also known as Dschubba, derived from the Arabic for 'forehead [of the scorpion]'.

OTHER BODIES

M7, or the Ptolemy Cluster, is an open cluster of bright, young blue stars easily visible with the naked eye. It's named after Claudius Ptolemy because his *Almagest* mentions it as a notable 'cloudy star' lying just outside the scorpion itself. It has a declination of −34.8°, making it the most southerly of all the 110 Messier objects.

METEOR SHOWERS

The Alpha Scorpiids are a medium-intensity meteor shower seen between 1 and 31 May, peaking around 16 May. A weaker shower is also associated with this constellation: the Omega Scorpiids peak on 2 June.

MYTHOLOGIES

The Babylonian star catalogue MUL.APIN of about 1000 BCE refers to the stars we now know as Scorpius as GIR.TAB, a 'scorpion'. This idea was then passed down to the ancient Greeks.

Greek-Egyptian astronomer Eratosthenes of Cyrene (276–194 BCE) and Greek poet Aratus (c. 315–240 BCE) both claimed that Artemis, the goddess of hunting, sent this scorpion to attack **60. Orion** after he wouldn't leave her alone. Eratosthenes also shared a different story, that Earth itself set the scorpion on Orion when he boasted that he could kill any creature. In another account, the scorpion was set upon **40. Hercules**.

INTERESTING FACTS

Scorpius is home to a number of intriguing exoplanets with characteristics ranging from extreme old age to potential habitability.

The planet PSR B1620-26 b, which is estimated to be 12.7 billion years old, is sometimes referred to as Methuselah, after an extremely long-lived man in the Bible. It's certainly a long-lived world; by contrast, the universe is approximately 13.7 billion years old. The planet has roughly twice the mass of Jupiter and orbits two stars at once, making it a 'circumbinary' planet. The two stars are known as PSR B1620-26 A and WD B1620-26, respectively.

Gliese 667 Cc is a super-Earth roughly four times the size of Earth. It orbits Gliese 667 C, a red dwarf star in a three-star system only 22 light years from Earth. According to astronomers from the Planetary Habitability Laboratory at the University of Puerto Rico at Arecibo, the planet is potentially habitable.

Notably, the same system also contains two more potentially habitable planets: Gliese 667 Ce and Gliese 667 Cf are both approximately 2.7 times the mass of Earth. Astronomers frequently define habitability as a rocky world close enough to its parent star for liquid water to exist on the surface (see '"Habitable" or "Goldilocks" Zones', page 61). Other factors, such as the composition of a planet's atmosphere and the variability of the host star, are also likely to be involved.

74. SCULPTOR

Pronounced: 'SKULP-ter'
Short: Scl
Brightest star: α Sculptoris (RA 0h 58m, Dec. −29°21')

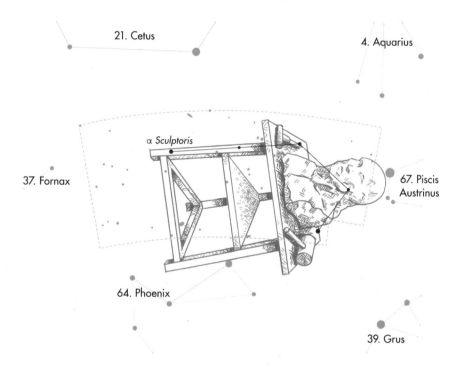

These stars had been noted for a long time, but Greek poet Aratus (c. 315–240 BCE) referred to this part of the sky simply as the 'waters' of very faint stars, of which only two were bright enough to be of any note.

WHEN AND WHERE TO OBSERVE
A small, faint constellation in the southern sky, Sculptor can be seen from latitudes south of +50°. It is most easily found by first identifying its neighbour, **4. Aquarius**, and it is most visible in November.

THE BRIGHTEST STARS
α Sculptoris is a giant B-type star. For all that it appears faint to us, it's thought to radiate about 1,500 times the energy of the Sun and to have a surface temperature of about 13,600K.

OTHER BODIES

There are no Messier objects in this constellation, but there are a few interesting deep-sky objects, which are extremely faint and can only be seen with large telescopes. The Sculptor Group of galaxies contains 13 individual members. The Sculptor Dwarf is an irregular galaxy located approximately 290,000 light years from Earth. NGC 253, also known as the Sculptor Galaxy, is a barred spiral galaxy. NGC 300 is a spiral galaxy 94,000 light years in diameter. NGC 7793 is a spiral galaxy 12.7 million light years away from Earth. The barred spiral galaxy NGC 613 is 67 million light years away.

METEOR SHOWERS

There are no significant meteor showers associated with this constellation.

MYTHOLOGIES

As one of the more recently identified constellations, and due to its small size and faint stars, there are no known mythologies associated with Sculptor.

INTERESTING FACTS

Sculptor is one of the 14 constellations first named by French astronomer Nicolas-Louis de Lacaille (see **2. Antlia**). In 1756, he called it *l'Atelier du Sculpteur*, the 'sculptor's studio', depicting it as a carved stone head on a table with various tools around it – but not including the artist involved in the sculpting. The sculpting equipment became the focus of the Latin name introduced in 1763, Apparatus Sculptoris. This has since been shortened to Sculptor, the literal meaning of which is the person doing the sculpting, rather than their tools or place of work.

There's been some adjustment to this and Lacaille's other constellations since he created them, but this is now the largest of his creations. However, it contains relatively few stars and none of them are particularly bright. That's especially evident when comparing this region of space to nearby bright Fomalhaut in **67. Piscis Austrinus**.

75. SCUTUM, THE 'SHIELD'

Pronounced: 'SKEW-tum'
Short: Sct
Brightest star: α Scuti (RA 18h 35m, Dec. −8°14')

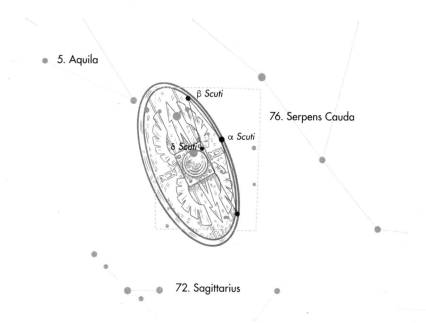

Of all the many conflicts in history, the Battle of Vienna in 1683 is the only one commemorated in a constellation. Polish astronomer Johannes Hevelius (see 45. Lacerta) created it in 1684 and named it Scutum Sobiescianum, Latin for 'shield of Sobieski', after the then king of Poland. It probably helped that, as well as being an effective military commander, Sobieski was also a great supporter of science – and of Hevelius.

WHEN AND WHERE TO OBSERVE
The four brightest stars of this constellation, though not very bright in themselves, make up a distinctive, narrow diamond shape that can be found due north of the Teapot in **72. Sagittarius**. Located in the southern hemisphere, it is visible at latitudes south of +74° from June to August.

THE BRIGHTEST STARS
This is a relatively faint constellation. Orange K-type α Scuti is the brightest star and has an apparent magnitude of just 3.83. β Scuti is a binary star: a yellow G-type and a fainter B-type that orbit one another every 2.3 years.

Blue-white F-type δ Scuti is a variable star that fluctuates between apparent magnitude of 4.6 and 4.79 every 4.65 hours. This star is also heading slowly in our direction. Calculations suggest that in about 1 million years it will pass within ten light years of the Sun, when it will be brighter than Sirius in **14. Canis Major** (currently the brightest star in the night sky).

OTHER BODIES

The Wild Duck Cluster, M11, is the most visible open cluster in Scutum. William Henry Smyth named it in 1844 because he thought it resembled a flock of ducks in flight. The cluster is 6,200 light years away from Earth and almost 200 light years across; it contains approximately 3,000 stars, making it a particularly rich cluster. It is estimated to be around 220 million years old, though some studies give older dates, and it is estimated to be 3,700–11,000 times as massive as the Sun.

METEOR SHOWERS

There are no significant meteor showers associated with Scutum.

MYTHOLOGIES

As one of the more recently identified constellations, and due to its small size and faint stars, there are no known mythologies associated with this constellation.

INTERESTING FACTS

In the summer of 1683, the vast Ottoman army lay siege to Vienna, capital city of Austria. This was the latest clash between the huge Ottoman Empire and the neighbouring (and also huge) Holy Roman Empire, and the latter responded with forces led by John III Sobieski, king of Poland. In September of that year, they won a victory over the Ottomans, and it's this victory that the constellation commemorates.

76. SERPENS, THE 'SERPENT'

Pronounced: 'SER-penz'
Short: Ser
Brightest star: α Serpentis or Unukalhai (RA 15h 44m, Dec. +6°25')

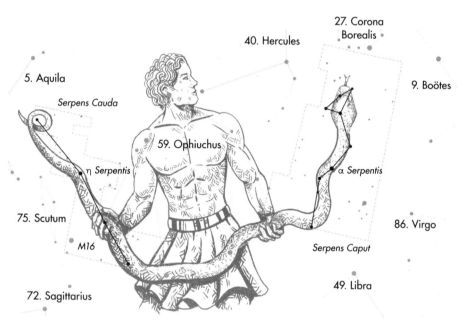

*Serpens is a very old constellation, dating back to Babylonian times. It is also unique among the 88 constellations because it comes in two disconnected parts, one on either side of **59. Ophiuchus**.*

WHEN AND WHERE TO OBSERVE
The two parts of Serpens are easily found on either side – and snaking round – **59. Ophiuchus** in the northern part of the sky. Serpens is best seen in the summer from the northern hemisphere, visible at latitudes between +80 and −80°.

THE BRIGHTEST STARS
Red giant α Serpentis is a K-type star known as Unukalhai, from the Arabic for 'serpent's neck'. A faint companion star is visible through a telescope, though we're not sure whether this is a binary system or whether these two components only seem close together when seen from Earth.
  Orange K-type η Serpentis is the brightest star in the tail section.

OTHER BODIES

M16, the Eagle Nebula, is a vast cloud of gas and young stars, with a dark region thought to resemble a swooping eagle. In 1995, the Hubble Space Telescope focused on part of this nebula and photographed three adjacent 'towers' of gas and dust in which new stars are being formed. The famous photograph is known as the 'Pillars of Creation'. (Astronomers also speak of such towers and pillars as 'elephant trunks'.)

METEOR SHOWERS

There are no significant meteor showers associated with this constellation.

MYTHOLOGIES

The Babylonians saw this pattern of stars as a horned serpent they called Bašmu. Since at least the time of Greek poet Aratus (c. 315–240 BCE), the same stars have been seen as a snake being carried by or encircling a central figure: Serpens Caput (the head) is on one side of him and Serpens Cauda (the tail) is on the other.

Aratus and other ancient Greeks referred to this snake constellation as Ophis. Eratosthenes of Cyrene (276–194 BCE), however, saw it as part of **59. Ophiuchus**, rather than a constellation of its own. Greek-Egyptian Claudius Ptolemy seems to have had it both ways: his *Almagest* of about 150 CE refers to Serpens as a distinct constellation but calls it Ophis Ophiouchou – the 'serpent of the serpent-bearer'. He may have named it this to differentiate it from two other snakes he recognised in the sky: **34. Draco** and **42. Hydra**.

INTERESTING FACTS

Whether Serpens is a distinct constellation or not was still being argued into the twentieth century; the decision to divide Serpens in two was first proposed in 1879 by American astronomer Benjamin Gould, but it took decades for this to be formally adopted.

77. SEXTANS, THE 'SEXTANT'

Pronounced: 'SECKS-tunz'
Short: Sex
Brightest star: α Sextantis (RA 10h 7m, Dec. −0°22')

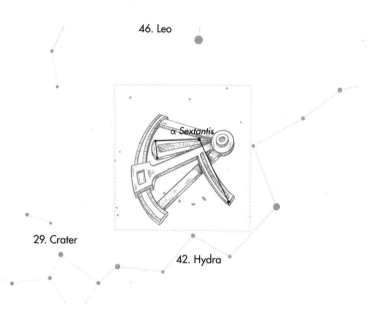

This small, faint constellation was created by Polish astronomers Elisabeth and Johannes Hevelius (see **45. Lacerta**) and included in the star catalogue that Elisabeth saw published in 1690, three years after her husband's death. They had in mind an astronomical instrument used for measuring angles, especially the angular positions of stars.

WHEN AND WHERE TO OBSERVE

This is an extremely faint and small constellation located south of the more distinctive **46. Leo** and straddling the celestial equator. From January to May, it is best seen from the southern hemisphere and is completely visible at latitudes ranging from +80 to −90°. Dark skies in April are your best bet for seeing it – but it isn't easy to spot!

THE BRIGHTEST STARS

This is a very faint constellation: A-type α Sextantis has an apparent magnitude of only 4.49. As with the other constellations created by Elisabeth and Johannes Hevelius, this might have been the point: even without a telescope, they were skilled enough to see them.

OTHER BODIES

Sextans contains no Messier objects, but it does have a few noteworthy deep-sky objects. NGC 3115 is a lenticular galaxy that appears almost edge on to our eyes. It is only 31.6 million light years away from Earth and is sometimes referred to as the Spindle Galaxy. It is also the nearest galaxy with a supermassive black hole at its core.

NGC 3169 and NGC 3166 are spiral galaxies that are only about 160,000 light years apart from one another and will eventually merge to form a larger galaxy. Sextans A and B are both irregular galaxies; Sextans B is one of the smallest known irregular galaxies with planetary nebulae.

METEOR SHOWERS

The Daytime Sextantids meteor shower occurs every year between 9 September and 9 October, peaking on 27 September. But how can we 'see' meteors during the day unless they're extremely bright? In fact, the ionised trail of a meteor reflects radio waves, which means that, by tuning a radio set to a special radio beacon, you can detect bursts of energy that indicate the passing of meteors during the day, at night and even in cloudy weather.

MYTHOLOGIES

As one of the more recently identified constellations, and due to its small size and faint stars, there are no known mythologies associated with Sextans.

INTERESTING FACTS

Large, wall-mounted 'mural instruments' have been used for measuring the positions of stars since ancient times. The wall would run parallel to the meridian (the imaginary straight line running between Earth's poles), and as a star was observed to pass over this line, the instrument would be used to measure its angular position – very basically, its height – in the sky, and the time was noted, too. This gave the star's declination (in degrees) and right ascension (in hours and minutes).

Early examples of these instruments tended to be mural quadrants, shaped like a quarter-circle and able to measure a 90° range of elevations. In about 994 CE, Persian astronomer Abu-Mahmud Khojandi built a huge mural sextant in his observatory near Ray in modern Iran. This measured a 60° range of elevations (hence the name, from the Latin for 'sixty'). That meant both declination and right ascension were being measured in ranges of 60, which made various calculations much easier.

Actually using Khojandi's sextant proved a little tricky, as it was 20m wide. Sultan Ulugh Beg (1394–1449) was a keen astronomer and had a mural sextant built at his observatory in Samarkand, in modern Uzbekistan, which was twice as wide as Khojandi's and had its own inbuilt staircase.

Soon sextants were being built that were no longer fixed to the position of the meridian. These more agile framed sextants could be used to measure the angular distance between any two given stars. Johannes Hevelius was particularly proud of his observations of even faint stars, which he conducted using a sextant and a quadrant – but not a telescope. When a fire destroyed his observatory and equipment in 1679, he felt the loss of the instruments keenly – and later created the constellation Sextans in their honour.

A much smaller and more portable version of the sextant was of great value in navigation and is often referred to as a nautical sextant because of its use by sailors. This superseded an earlier device commemorated in another constellation, **58. Octans**.

78. TAURUS, THE 'BULL'

Pronounced: 'TORE-us'
Short: Tau
Brightest star: α Tauri or Aldebaran (RA 4h 35m, Dec. +16°30')

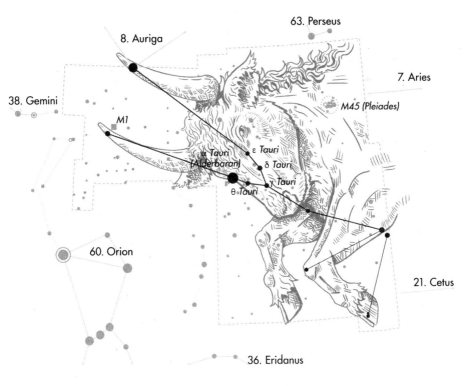

Some claim that this well-known constellation from the zodiac is the oldest-recognised constellation still in use today. The Lascaux caves in south-west France contain more than 600 wall paintings thought to date from about 15,000 BCE. Among these paintings is a particular section that shows a bull, a half-man, half-bird figure and a bird on a stick, with a pattern of spots to one side. It's been argued that these strange figures match the positions of well-known stars: the bull figure is modern Taurus, the pattern of spots the star cluster we now call M45, the Pleiades.

WHEN AND WHERE TO OBSERVE
This large, prominent constellation with its distinctive 'V' shape is very easily found: the three bright stars in the belt of **60. Orion** point towards it – and the Pleiades are just beyond. It's a winter constellation in the northern sky, visible at latitudes between +90 and −65° through January.

THE BRIGHTEST STARS

α Tauri is a bright red K-type star known as Aldebaran, from the Arabic for 'follower', because it seems to follow the Pleiades.

γ Tauri, ε Tauri, θ Tauri and the main component of δ Tauri are all part of an open cluster of stars known as the Hyades, some 153 light years from Earth. The cluster takes its name from five sisters in Greek mythology, the half-sisters of the nearby Pleiades. The cluster has been known by this name since at least 750 BCE, as it's mentioned in Homer's *Iliad*.

OTHER BODIES

M45, the Pleiades, is a very well-known open cluster, full of hot, bright blue stars that formed within the last 100 million years – so they are relatively young, as stars go. In Greek mythology, these were the seven daughters of the nymph Pleione. The cluster is still often known as the Seven Sisters, though more than seven stars can be seen in it, even with the naked eye. Some think the mythology took the name from the stars, not the other way round, and that the name Pleiades originally derived from a word for 'sailing', as these bright stars were used in navigation.

A telescope is needed to see another notable object. M1 was the first object listed in the catalogue of unusual objects made by French astronomer Charles Messier (1730–1817). It's also known as the Crab Nebula after a (not very good) illustration from 1844 made it look a bit crab-like.

METEOR SHOWERS

The Taurid meteor shower is caused by debris left behind by Comet Encke. This comet stream is very spread out and dispersed, which is why the shower is relatively long. It is also why we see two separate segments of the shower: the South Taurids (10 September to 20 November) and the North Taurids (20 October to 10 December). As two fragments of the same debris cloud, they are similar in density.

MYTHOLOGIES

Whatever the merits of the cave paintings mentioned above, there's evidence that this bull-shaped constellation was recognised by the Sumerians as far back as 3200 BCE (see **73. Scorpius**). In the MUL.APIN star catalogue of about 1000 BCE, these stars are listed as GU.AN.NA, the 'bull of heaven' – one such bull was sent by the goddess Ishtar to kill the hero Gilgamesh in Babylonian mythology.

A number of neighbouring cultures worshipped bulls, too. In Egypt, the red bull Apis (not to be confused with **3. Apus**) was thought to be a form taken by the god Osiris. The Israelites worshipped a Golden Calf, as detailed in the Bible. On Crete, there were bull-leaping ceremonies and evidence of bull worship, and in Assyria and India we find evidence of religious devotion to bulls. There are numerous theories about how these different cultures and their interest in bulls might have been interlinked.

The Greeks applied various meanings to the bull-shaped constellation. It was the Cretan Bull fought by **40. Hercules**, and father of the half-bull, half-human Minotaur. Or it was Zeus in disguise. Or it was someone Zeus had transformed into a bull, such as human princess Io. Perhaps having so many different explanations is itself revealing: the constellation has been known for so long by so many different cultures that there are a great many stories associated with it.

INTERESTING FACTS

We now know that M1, the Crab Nebula, is what's left of an exploded star. In fact, it is the first object we ever recognised as being the result of a supernova. What's more, by measuring the rate at which this cloud of gas is expanding, we are able to calculate when it formed, and match that to historical records of unusual sightings in the night sky.

In 1054, observers in China, Japan and Baghdad all recorded the appearance of a bright new 'guest star', visible from late April or early May of that year and getting steadily brighter until early July, when it was brighter than Venus and all the stars in the night sky. Then it faded away and wasn't seen again until the 1700s, when Messier and other astronomers using telescopes happened upon the faint remnant of M1.

79. TELESCOPIUM, THE 'TELESCOPE'

Pronounced: 'tel-iss-KOP-ee-um'
Short: Tel
Brightest star: α Telescopii (RA 18h 26m, Dec. −45°58')

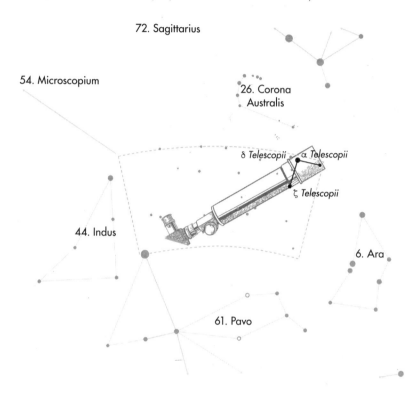

72. Sagittarius

54. Microscopium

26. Corona
Australis

δ Telescopii   α Telescopii

ζ Telescopii

44. Indus

6. Ara

61. Pavo

This small, faint constellation in the southern skies is one of the 14 constellations first named by French astronomer Nicolas-Louis de Lacaille (see **2. Antlia**). *We tend to think of telescopes as tubes, but Lacaille probably had a different sort in mind for this pattern of stars, as we shall see.*

WHEN AND WHERE TO OBSERVE
Telescopium is visible from latitudes south of +40°. Though it is very small and faint, it can be located in dark skies thanks to the distinctive curl of neighbouring **26. Corona Australis**, and it is best viewed in August.

THE BRIGHTEST STARS
Blue-white B-type α Telescopii is the brightest star in Telescopium. Greek-Egyptian astronomer Claudius Ptolemy observed this star, but in his *Almagest* of about 150 CE he considered it part of **26. Corona Australis**.

The star is just over five times the mass and three times the diameter of the Sun, but it is radiating almost 800 times the energy.

There is no β Telescopii, as the star once identified as such has been reassigned: it is now η Sagittarii in **72. Sagittarius**. Orange K-type ζ Telescopii is now the second brightest star here. δ Telescopii is in fact two separate blue-white B-types that only appear close together as seen from Earth.

OTHER BODIES

There are no Messier objects in Telescopium, and while there are a few interesting deep-sky objects, these are extremely faint and can only be seen with large telescopes. The Telescopium Group is a group of 12 galaxies located approximately 120 light years from Earth. NGC 6861 is a stunning lenticular galaxy (a cross between a spiral and an elliptical galaxy) with deep dust lanes. NGC 6584 is a globular star cluster made up of thousands of stars.

METEOR SHOWERS

There are no significant meteor showers associated with this constellation.

MYTHOLOGIES

As one of the more recently identified constellations, and due to its small size and faint stars, there are no known mythologies associated with Telescopium.

INTERESTING FACTS

Telescopes were a new invention when, in 1609, Italian astronomer Galileo Galilei built one for himself and trained it on the sky. In his book *Starry Messenger*, published the following year, Galileo included what we think are the first observations of any stars seen through a telescope: **60. Orion** and the Pleiades in **78. Taurus**.

There was soon a craze for telescopes and stargazing across Europe, and the next few decades saw marked improvements in telescope technology. The wider the objective lens, the more light it gathered and the more detail could be seen. However, the wider the objective lens, the further from it the eyepiece needed to be to get the image in focus. This focal length is equal to the square of the objective diameter: if the objective lens was made five times wider, the focal length had to be made 25 times longer!

Long focal lengths had other advantages, such as reducing the appearance of rainbow-like halos known as chromatic aberration. But they also made telescopes huge and unwieldy and prone to being blown about by the wind. This made it difficult to make accurate observations of stars.

Then, around 1675, Dutch brothers Christiaan and Constantijn Huygens became the first astronomers to use a new kind of telescope that did away with the cumbersome tube.

In an aerial telescope, the objective lens is fixed with a swivelling joint to some high vantage point, such as the top of a pole or the side of a building. A string or rod leads from this objective lens down to the astronomer on the ground, ready with an eyepiece. By keeping the string taut (or using the rod), the astronomer maintains the correct focal length – or distance from the objective lens – to produce a magnified image. The tubeless apparatus is relatively lightweight and easy to manoeuvre, so the astronomer can quickly move it round to look at different parts of the sky.

This device is what Lacaille seemed to see in this pattern of stars: a structure (housing the objective lens) and a trailing string. He called it *le Telescope* in his star map of 1756, and this was translated into the Latin *Telescopium* for the 1763 edition.

But it seems other astronomers saw an older kind of telescope in these stars. In the decades that followed, it was variously named Tubus Telescopium ('telescope tube') and Tubus Astronomicus ('astronomical tube'). The Telescopium of today is also much smaller than the one Lacaille created; several of the stars he saw as part of the aerial telescope have been reassigned to other constellations.

80. TRIANGULUM, THE 'TRIANGLE'

Pronounced: 'try-AN-gyoo-lum'
Short: Tri
Brightest star: β Trianguli (RA 2h 9m, Dec. +34°59')

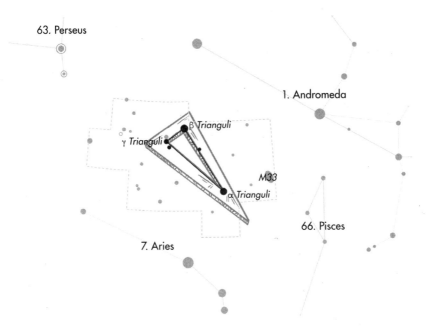

This simple pattern of three stars has been recognised for a very long time.
The Babylonian star catalogue written in about 1000 BCE is known by the
name of the first constellation it listed: MUL.APIN. The word means 'plough',
and the Babylonians saw a huge plough shape made up of stars from what
we now recognise as Triangulum, **1. Andromeda** and **18. Cassiopeia**.
(This Babylonian plough should not be confused with the smaller Plough or
Big Dipper seen in **83. Ursa Major.**)

WHEN AND WHERE TO OBSERVE
A small, relatively faint constellation in the northern skies, the brightest
stars here nevertheless make up a distinct triangle shape that borders the
well-known **1. Andromeda**, **7. Aries** and **66. Pisces**. Triangulum is most
prominent in December.

THE BRIGHTEST STARS
White β Trianguli is the brightest star in the constellation, a binary whose
main component is a giant A-type. The two stars orbit one another

every 31.4 days, at a distance of less than five times that between Earth and the Sun.

Yellow-white α Trianguli is also a binary star; its brighter component is known as Mothallah, from an Arabic phrase meaning 'head of the triangle'. Together, the two components are classed as an F-type and they orbit one another every 1.7 days.

White A-type γ Trianguli is orbited by a disc of dust and debris, which we can detect because it is heated by the star's energy to about 75K (or −198°C).

OTHER BODIES

M33, the Triangulum Galaxy, is just visible to the naked eye in dark skies. It's a spiral galaxy some 3 million light years from Earth, making it one of the closest members of our local group of galaxies. At 61,000 light years in diameter, it's much smaller than our Milky Way or even the nearby Andromeda Galaxy, which we think M33 might be orbiting. However, because M33 is angled face on towards us, the light of its individual stars is very spread out (or diffuse). A telescope with a wide field of view is needed to see its structure in any detail, but it is well worth the effort. Its fine structure, when visible, is why M33 is also known as the Pinwheel Galaxy.

METEOR SHOWERS

There are no significant meteor showers associated with this constellation.

MYTHOLOGIES

As one of the more recently identified constellations, and due to its small size and faint stars, there are no known mythologies associated with this constellation.

INTERESTING FACTS

While it seems that the ancient Greeks inherited several constellations from the Babylonians, the Babylonian plough was not one of them. To the Greeks, this triangular pattern of stars was Deltoton, after the Greek letter delta or Δ. Greek-Egyptian astronomer Eratosthenes of Cyrene (276–194 BCE) thought Deltoton represented the Nile Delta, a distinctive, triangle-shaped landform where the river Nile meets the Mediterranean.

Roman author Gaius Julius Hyginus (c. 64 BCE–17 CE) saw the same constellation as the triangular island of Sicily, placed in the sky by the goddess Ceres, who also gave the island its name. Greek-Egyptian Claudius Ptolemy, writing in about 150 CE, knew the constellation simply as Trigonon – the 'triangle' – when he included it in *Almagest*.

81. TRIANGULUM AUSTRALE, THE 'SOUTHERN TRIANGLE'

Pronounced: 'try-AN-gyoo-lum aw-STRAL-ee'
Short: TrA
Brightest star: α Trianguli Australis or Atria (RA 16h 48m, Dec. −69°1')

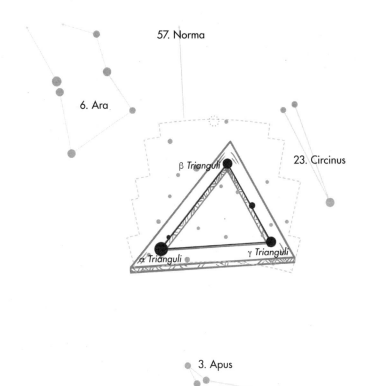

Triangulum Australe is the sixth smallest constellation and the smallest of the 15 constellations named by the Dutch astronomer Petrus Plancius and designed to fill in the blanks in maps of the southern sky.

WHEN AND WHERE TO OBSERVE

Visible from latitudes south of +25°, the three brightest stars of this constellation form an almost equilateral triangle. This constellation sits in a relatively dark part of the sky, due west of the bright α Centauri and β Centauri in **19. Centaurus**. In the southern hemisphere, this is a winter constellation, most visible in July.

THE BRIGHTEST STARS

Bright orange K-type α Trianguli Australis is known as Atria, a shorted version of its Bayer designation. There is some evidence that this is a binary star in orbit with a young and magnetically active G-type star.

β Trianguli Australis is an F-type star that evidence from the Spitzer Space Telescope suggests may have a disc of dust and debris in orbit around it. We're not sure if the star visible close by is part of a binary system or just appears close as seen from Earth.

OTHER BODIES

White A-type γ Trianguli Australis completes the Southern Triangle. We've detected unusually high levels of infrared radiation from this star, which we think is evidence of another dusty disc.

METEOR SHOWERS

There are no significant meteor showers associated with this constellation.

MYTHOLOGIES

As one of the more recently identified constellations, and due to its small size and faint stars, there are no known mythologies associated with this constellation.

INTERESTING FACTS

The Arctic Circle is an imaginary line of latitude where the Sun is just visible at noon on the shortest day of the year. It's currently 66°33' north of the equator. The word 'Arctic' derives from an ancient Greek phrase meaning 'place of (or close to) the Bear'. Although there are polar bears in the Arctic, we don't think the Greeks ventured this far; the name referred to the north being closer to the Great Bear constellation, **83. Ursa Major.**

Greek mapmaker Marinus of Tyre (c. 70–130 CE), whose work was a great influence on Claudius Ptolemy, coined the term 'Antarctic', meaning 'opposite of the Arctic', to refer to the most southerly part of Earth. To Marinus, this was largely uncharted territory.

In the early 1500s, European sailors began to explore the southern seas and they recorded observations of the stars they saw there. Amerigo Vespucci (1451–1512) is thought to have included a triangular constellation in his now lost star catalogue, but we're not sure which stars these were. In 1589, Dutch mapmaker Petrus Plancius (see **30. Crux**) named a pattern of three southern stars Triangulus Antarcticus, the 'triangle of the Antarctic', in contrast to the triangle of the north, **80. Triangulum.**

However, Plancius placed this new constellation in the wrong part of the sky. In 1603, German mapmaker Johann Bayer (1572–1625) published his star atlas *Uranometria* with these stars positioned more accurately and labelled Triangulum Australe, the Latin for 'Southern Triangle'.

82. TUCANA, THE 'TOUCAN'

Pronounced: 'too-KAH-nuh'
Short: Tuc
Brightest star: α Tucanae (RA 22h 18m, Dec. −60°15')

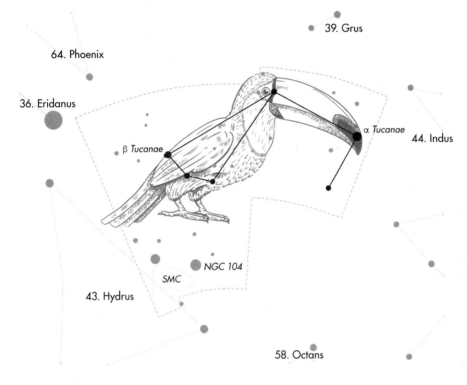

*Tucana is one of the four Southern Birds – the others are **39. Grus, 61. Pavo** and **64. Phoenix** – created at the same time and based, we think, on real birds seen by Dutch sailors on their journeys round the world.*

WHEN AND WHERE TO OBSERVE
Another small, faint constellation in the southern sky, Tucana is visible from latitudes south of +25°. It can be found due east of bright α Eridani, at the end of the river **36. Eridanus**, and is most prominent in November.

THE BRIGHTEST STARS
α Tucanae is a binary star, its main component being an orange K-type orbiting a fainter companion every 11.5 years. β Tucanae is a system of six stars, the two brightest ones being a blue-white B-type and a white A-type.

OTHER BODIES

The Small Magellanic Cloud is an irregular and relatively small galaxy with a central, bar-like structure. It's about 200,000 light years away, so among the closest neighbours of our own Milky Way, and it's one of the furthest objects that can be seen with the naked eye (see **33. Dorado** for details of the Large Magellanic Cloud).

A bright, fuzzy object that looks, from Earth, quite close to the Small Magellanic Cloud is really a bright globular cluster of stars classified as NGC 104.

METEOR SHOWERS

There are no significant meteor showers associated with this constellation.

MYTHOLOGIES

As one of the more recently identified constellations, and due to its small size and faint stars, there are no known mythologies associated with this constellation.

INTERESTING FACTS

This is another of the 15 constellations first identified by Dutch mapmaker Petrus Plancius (see **30. Crux**), based on observations made by sailors Pieter Dirkszoon Keyser and Frederick de Houtman (see **22. Chamaeleon**). It first appeared on the celestial globe made by Plancius and his colleague Jodocus Hondius in 1598.

More than 40 species of toucan are known, and they have distinctive, often colourful bills. Toucans are found throughout Central America, from southern Mexico to northern Argentina. The name comes from *tukana*, the name for these birds used by the Tupi people of southern Brazil, who shared the word with Portuguese sailors passing through the region, who then passed it down to us.

83. URSA MAJOR, THE 'GREAT BEAR'

Pronounced: 'ER-suh MAY-jur'
Short: UMa
Brightest star: ε Ursae Majoris or Alioth (RA 12h 54m, Dec. +55°57')

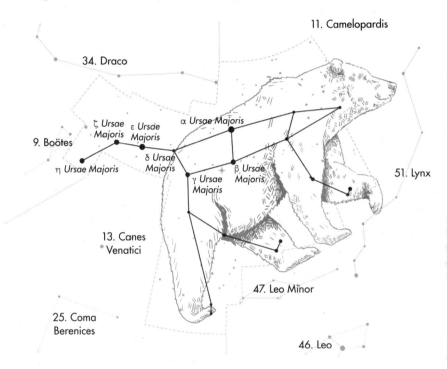

The seven brightest stars of this constellation make up one of the most easily recognised asterisms in the sky, the so-called Plough or Big Dipper. To the Babylonians, this was a wagon or 'wain', recognised by many later cultures who offered their own interpretations as to whose wain it might be (see 'What Are Constellations?', page 2).

WHEN AND WHERE TO OBSERVE

Large, bright and distinctive, Ursa Major is visible at latitudes north of −30°. What's more, from north of +30° it is a permanent fixture in the night sky, circling round the pole star in neighbouring **84. Ursa Minor**, but never setting behind the horizon. It is our constant companion.

THE BRIGHTEST STARS

ε Ursae Majoris is the brightest star in the constellation, an A-type that may be a binary or a multiple star system and that might have at least one large exoplanet in orbit around it – we're not sure! The name Alioth comes from an Arabic phrase meaning 'sheep's tail'.

Yellow α Ursae Majoris is a binary star, with another binary star system relatively close by making a quadruple system. It's known as Dubhe, from the Arabic for 'bear', suggesting that the same culture saw this constellation as both a sheep and bear.

η Ursae Majoris is a blue-white B-type star. It's also known as Alkaid, from yet another interpretation of this group of stars. The word comes from an Arabic phrase meaning 'leader of the mourning daughters': three daughters lead the 'box' of a funeral bier.

OTHER BODIES

ζ Ursae Majoris is also known as Mizar, from the Arabic for 'covering' or 'apron'. Seen with the naked eye, it has a close companion, Alcor, meaning 'the neglected one', in comparison to the brighter Mizar. In fact, the two stars only appear close together when seen from Earth and are not otherwise connected.

In 1617, Italian astronomer Galileo Galilei looked at Mizar through a telescope and saw it had a companion, making this the first ever binary star detected using a telescope. (We think another Italian astronomer, Benedetto Castelli, saw it first and told Galileo, who was the first to record it.)

The two components of Mizar are often known as Mizar A and Mizar B. We now know that they are both themselves binary stars, so Mizar is a quadruple system. The two stars in Mizar A are each some 35 times as bright as the Sun, and they take a little more than 22.5 days to orbit one another. That length of time was first calculated by Antonia Maury (1866–1952), an American astronomer who had to battle to get proper credit for her brilliant work, which kept being published under the names of men she worked with!

β Ursae Majoris is an A-type star also known as Merak, from the Arabic for 'loins of the bear'. A line drawn from this star through α Ursae Majoris and continuing on leads to the pole star, Polaris, the brightest star in **84. Ursa Minor**. In the other direction, the same stars point towards the 'sickle' or upside-down question mark in **46. Leo**.

γ Ursae Majoris is a binary star with an A-type main component and a K-type companion. Its name Phecda comes from the Arabic for 'thigh [of the bear]'.

δ Ursae Majoris is a white A-type star known as Megrez, from the Arabic for 'base [of the bear's tail]'. We can follow the arc of the bear's tail to red giant Arcturus in **9. Boötes** (and then continue on to 'speed to Spica' in **86. Virgo**).

This constellation is also chock-a-block with brilliant objects to observe, including seven Messier objects. Among these is a double star, a planetary nebula, an irregular galaxy and four spiral galaxies.

M81, also known as Bode's Galaxy, is a large spiral galaxy 11.8 million light years away from Earth. Its brightness makes it a popular target for amateur astronomers using small telescopes. M82, the Cigar Galaxy, is a cigar-shaped, edge-on starburst galaxy. M97, also known as the Owl Nebula, is a planetary nebula with two dark patches that resemble the eyes of an owl. M101, also known as the Pinwheel Galaxy, is a magnificent face-on spiral galaxy with bright spiral arms and dark dust lanes.

METEOR SHOWERS
The Ursa Majorids are a relatively minor meteor shower, with fewer than one meteor an hour seen even at the peak on 15 October. Much more notable are the Leonid-Ursids that can be seen in the week before Christmas with anything up to ten meteors per hour.

MYTHOLOGIES
Various explanations have been given for how these stars came to be seen as a bear. Bears are commonly featured in cave paintings and other early artwork, and it has been suggested that – as with **78. Taurus** – some of these may represent patterns of stars. However, the bear in Ursa Major has a notably long tail, which real bears do not. So why were these stars not interpreted as some other large, long-tailed animal?

In the *Iliad*, written about 750 BCE, Homer refers to the 'Great Bear known as the Wain', suggesting that the ancient Greeks recognised two ways of seeing the same pattern of stars. One, the Wain, was surely inherited from the Babylonians, so the Great Bear had to come from somewhere else.

At the time Homer was writing, trade routes reached from Greece to India and beyond. There is evidence of ideas being shared and spreading between the two cultures. In the Vedic tradition in what is now India, these same seven bright stars are the Saptarishi or 'seven wise people', but the last part of that word is very like *riksha*, meaning both 'star' and 'bear'. It's been suggested that somewhere along the trade route, a small mistranslation of the Indian word created the bear constellation. Some even think that the Latin word for bear – *Ursa* – might derive from that Indian word.

Whatever the case, the ancient Greeks weren't the only people to see these stars as a bear. To both the Cherokee and Iroquois people of North America, what we see as the bear's long tail is nothing of the sort; these stars are human hunters following the bear. To the Wasco-Wishram people, there are two bears and five wolves in these stars.

The Greeks were still using two names for the constellation by the time of Greek poet Aratus (c. 315–240 BCE), who referred to these stars as the 'wagon-bear'. He also called them Helike or 'twister' because they circle the north celestial pole. Aratus thought this and the smaller **84. Ursa Minor** were the bears that nursed the god Zeus as a child, and other Greek writers supplied names for these nurse-bears: Adrasteia and Ida. But Ursa Major was also associated with the Greek myth of Callisto, a princess transformed into a bear and then unwittingly hunted by her own son.

INTERESTING FACTS

In December 1995, the Hubble Space Telescope took 342 separate long-exposure photographs of a tiny, tiny part of the sky in Ursa Major. These were compiled to produce the Hubble Deep Field image, which looked further into space than ever before. For all that it was focused on a tiny field of view, the image is densely speckled with some 3,000 objects large and small, in a rich variety of colours. They're mostly all distant galaxies, each one containing thousands of millions of stars – powerful evidence of how much more is out there for us to discover!

84. URSA MINOR, THE 'LITTLE BEAR'

Pronounced: 'ER-suh MY-nur'
Short: UMi
Brightest star: α Ursae Minoris or Polaris, the North Star (RA 2h 31m, Dec. +89°15')

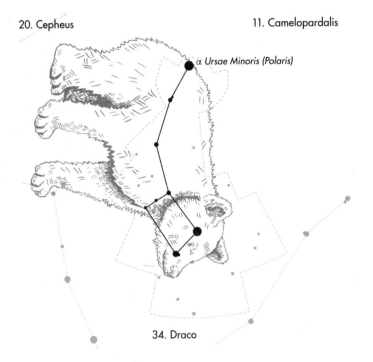

20. Cepheus

11. Camelopardalis

α Ursae Minoris (Polaris)

34. Draco

*The pattern of stars here looks like a smaller, fainter mirror image of the well-known Plough or Big Dipper in nearby **83. Ursa Major**. To the Babylonians, these were both wagons or 'wains' of different sizes.*

WHEN AND WHERE TO OBSERVE

A line from the two brightest stars of neighbouring, bright **83. Ursa Major**, from β Ursae Majoris, passing through α Ursae Majoris on the outside edge of the Plough, points towards α Ursae Minoris or Polaris. Ursa Minor can be seen on any clear night from anywhere in the northern hemisphere of Earth, and for some of the year even as far south as −10° latitude.

THE BRIGHTEST STARS

Yellow-white α Ursae Minoris, also known as Polaris or the North Star, is at +90° declination, so directly above Earth's North Pole. Effectively, our planet rotates underneath it. This is very useful for navigators: the direction in which Polaris lies lowest in the sky is due north, and the elevation of Polaris above the northern horizon is equal to the observer's latitude. The name is a shortened form of *Stellar Polaris*, Latin for 'polar star'. Polaris is actually a triple system of F-types.

OTHER BODIES

β Ursae Minoris is an orange-coloured red giant K-type star also known as Kochab, though we're not sure why – it may be an abbreviation of an Arabic or Hebrew phrase. The colour is a sign of relatively low temperature. It is much cooler than our own Sun, and yet it is radiating 390 times as much energy. The discrepancy is explained by Kochab being so much larger, with a diameter 42 times greater than that of the Sun.

METEOR SHOWERS

The Ursids meteor shower occurs every year between 17 and 24 December, usually reaching its peak around 23 December, when observers may be able to see up to ten meteors per hour.

MYTHOLOGIES

Greek poet Aratus (c. 315–240 BCE) knew these stars as Kynosaura, which means 'dog's tail', yet there's no corresponding dog-shaped constellation nearby. He also claimed that this was one of the two bears that nursed Greek god Zeus as a child. It may be that the constellation was considered to resemble a bear, but one with a long tail like a dog's. That name, and the usefulness of this constellation in navigation, is the root of the English word 'cynosure' or 'guiding star'.

By the time Greek-Egyptian Claudius Ptolemy included the constellation in *Almagest* in about 150 CE, it was named Arktos Mikra, the small bear. Ursa Minor is the Latin form of this name.

INTERESTING FACTS

In the *Iliad*, written by Greek poet Homer in about 750 BCE, there's a reference to the 'Great Bear known as the Wain' but no mention of a second smaller wain or bear. Thales of Miletus (c. 624–c. 548 BCE) is thought to have made the first reference to a 'little bear' constellation, though his work is lost and we only know what he wrote from surviving accounts by later writers. Thales called this constellation – or thought it should be called – the Phoenician Bear because Phoenician sailors from (roughly) modern Lebanon used these stars in navigation. Greek-Egyptian astronomer Eratosthenes of Cyrene (276–194 BCE) knew this constellation simply as the Phoenician.

85. VELA, THE 'SAILS'

Pronounced: 'VEE-luh'
Short: Vel
Brightest star: γ Velorum or Suhail al Muhlif or Regor (RA 8h 9m, Dec. −47°20')

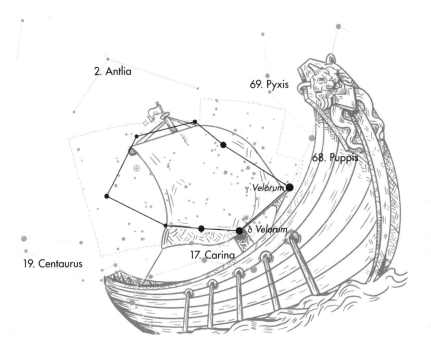

This was once part of the great boat constellation Argo Navis, broken up into more manageable pieces by French astronomer Nicolas-Louis de Lacaille in 1752 (see **17. Carina**). Vela is the boat's sails, **17. Carina** is the keel and **68. Puppis** the stern.

WHEN AND WHERE TO OBSERVE
The brightest stars in this constellation form a sail or shield-shaped pattern in the southern sky, due east of **19. Centaurus**. This is visible at latitudes south of +30° and is best seen from the southern hemisphere in March. Knowing the other parts of the Argo Navis will also help you to identify this constellation.

THE BRIGHTEST STARS
Lacaille kept the Bayer designations given to Argo Navis as a whole, so the brightest star in Vela is γ Velorum. This is a system of at least four stars,

the brightest component being a binary star made up of a blue supergiant O-type and what's called a Wolf–Rayet star.

These unusual stars range in temperature from 20,000 to 210,000K, hotter than almost all other types of stars. The one here is thought to be 57,000K and radiating energy at 170,000 times the rate of the Sun – while being just six times the diameter and nine times the mass! We think this is the result of the star having burnt through its hydrogen fuel and now being in the transformative process that will lead to a supernova.

The O-type star is actually brighter than the Wolf–Rayet star, radiating 280,000 times the energy of the Sun. It is bigger, too: 30 times as massive as the Sun and 17 times its diameter. Yet its surface temperature is cooler than the Wolf–Rayet companion, at a still very hot 35,000K. They are quite a pair!

OTHER BODIES
δ Velorum is a triple system composed of two A-types and an F-type. It is known as Alsephina, from the Arabic for 'ship', which is the name by which Argo Navis as a whole was once known.

METEOR SHOWERS
There are no significant meteor showers associated with this constellation.

MYTHOLOGIES
As one of the more recently identified constellations, and due to its small size and faint stars, there are no known mythologies associated with this constellation.

INTERESTING FACTS
A number of star systems within this constellation have been found to have exoplanets. HD 85390 is an orange dwarf located approximately 111 light years away, with a planet around 30 times the mass of Earth orbiting it every 788 days. HD 93385 is a Sun-like star with two super-Earths orbiting it that have periods of 13 and 46 days and masses of 8.3 and 10.1 times that of Earth, respectively.

HD 75289 is another Sun-like star with a hot Jupiter planet that orbits it every 3.5 days at a distance about 21 times less than that from Earth to the Sun. WASP-19 sits 815 light years away from Earth and has a hot Jupiter-like planet with an orbit of only 0.7 days! HD 73526 is also a Sun-like star, which has two planets with orbital periods of 187 and 377 days, respectively.

86. VIRGO, THE 'MAIDEN'

Pronounced: 'VER-go'
Short: Vir
Brightest star: α Virginis or Spica (RA 13h 25m, Dec. –11°9')

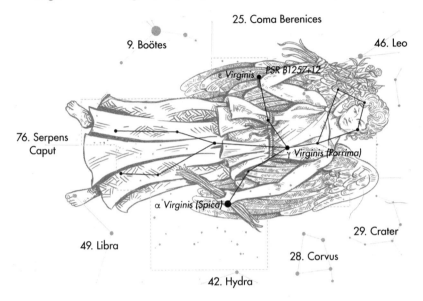

Virgo will be familiar to many as it is one of the zodiac constellations. Its name comes from the Latin word for 'maiden', and it is one of the 48 constellations catalogued by Greek-Egyptian astronomer Claudius Ptolemy in the second century. It is the second largest constellation in the sky (after **42. Hydra***) and the largest in the zodiac.*

WHEN AND WHERE TO OBSERVE
Visible between latitudes +80 and –80°, Virgo can be seen in spring and summer in the northern hemisphere and in autumn and winter in the southern hemisphere. This large constellation is easily found in the northern hemisphere by using two other prominent constellations. The handle of the Plough or the tail of **83. Ursa Major** can be used to 'arc to Arcturus', the brightest star in **9. Boötes**. We then can carry straight on and 'speed to Spica', one of the stars that make up this human stick-figure pattern. Virgo is also located between **46. Leo** to the west and **49. Libra** to the east.

THE BRIGHTEST STARS
α Virginis or Spica is a binary star; its B-type components are so close together that they orbit one another every four days. (For the origin of the name Spica, see 'Mythologies', below.)

γ Virginis, the second brightest star in the constellation, is also a binary system, this time composed of two F-type stars. It's known as Porrima, a Roman goddess of prophecy.

The third brightest star, ε Virginis, is a single G-type with a mass 2.6 times that of the Sun. Its name, Vindemiatrix, means 'the woman harvesting grapes' because, to the ancient Greeks, it was first visible in August, when it marked the start of the wine harvest.

OTHER BODIES

This region of space is rich in galaxies: there are more than 13,000 of them between Virgo and neighbouring **25. Coma Berenices**. These are best explored by telescope.

METEOR SHOWERS

The Virginids are a series of weak meteor showers, collectively lasting from late January into early May, and peaking in March and April. With on average of only one or two meteors per hour, they are easy to miss.

MYTHOLOGIES

In the Babylonian star catalogue MUL.APIN from about 1000 BCE, this pattern of stars is a furrow in the ground, with the bright star α Virginis representing an ear of wheat – a *Spica* in Latin, which is still the name given to this star.

The Babylonians associated this furrow with Shala, goddess of farming. The ancient Greeks seem to have adopted some element of this – they saw the same ear of wheat held by a young woman, who is sometimes depicted with wings. This woman they knew as Parthenos, meaning 'maiden'.

Greek poet Aratus (c. 315–240 BCE) thought this constellation was Dike, a goddess who lived among humans until, sickened by their violence and war, she flew off into the night sky. Later writers suggested that these stars were Demeter, the Greek goddess of farming – surely related to the Babylonian Shala. Demeter's daughter Persephone was married to Hades, god of the underworld, and spent half the year with him there. In the months Persephone was away, her mother Demeter mourned her, and nothing would grow on the land. In the months Persephone was reunited with her mother, the ground was fertile for farming. This was how the ancient Greeks explained the changing seasons.

INTERESTING FACTS

The first three exoplanets ever discovered are in Virgo, orbiting a pulsar (see **88. Vulpecula**) known as PSR B1257+12. The strangeness of this system is reflected in the names given to the star – Lich, a word for 'undead creature' – and its planets: Draugr (an undead creature from Norse mythology), Poltergeist (a ghost) and Phobetor (a figure from Greek mythology who appears in people's dreams).

87. VOLANS, THE 'FLYING FISH'

Pronounced: 'VOH-lanz'
Short: Vol
Brightest star: γ^2 Volantis (RA 7h 8m, Dec. −70°29')

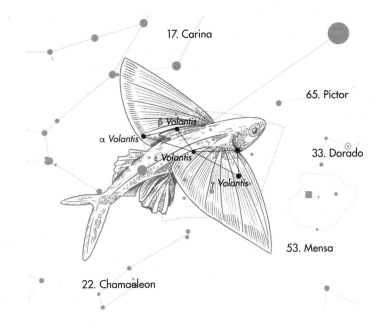

This constellation was originally called Vliegendenvis or 'flying fish' on the globe made in 1598 by Dutch mapmaker Petrus Plancius (see **30. Crux**) and Jodocus Hondius. In 1603, the German mapmaker Johann Bayer labelled the same constellation in Latin: Piscis Volans. This was later shortened to Volans, which helped to avoid confusion with **66. Pisces** and **67. Piscis Austrinus**.

WHEN AND WHERE TO OBSERVE

This small, faint constellation can be found due south of – or swimming beside – the keel of **17. Carina**. It is located in the southern hemisphere and is completely visible at latitudes south of +15° from December to February. A tip for spotting Volans: its brightest stars make up a pair of low, wide triangles, almost a bowtie shape.

THE BRIGHTEST STARS

γ Volantis is a binary system comprised of orange K-type γ^2 Volantis and a smaller F-type. α Volantis is another binary, its main component being an

A-type. Orange K-type β Volantis is a single star. ε Volantis is a quadruple system comprising two B-types and two A-types.

OTHER BODIES

The Lindsay–Shapley ring, also known as AM0644-741, is a ring galaxy located 300 million light years from Earth. In 1960, American astronomer Eric Lindsay, a student under Harlow Shapley, was the first to publish an account of this ring, located near the Large Magellanic Cloud. The unusual shape of this galaxy is thought to be the result of a collision with another galaxy, many millions of years ago. There are two main components: the blue ring is 150,000 light years in diameter and was formed when a shock-wave from the collision created a lot of hot, blue stars; the yellow core is an amalgamation of the original stars' cores.

Also located in this constellation is NGC 2442, an intermediate spiral galaxy that sits 50 million light years from Earth.

METEOR SHOWERS

There are no significant meteor showers associated with this constellation.

MYTHOLOGIES

As one of the more recently identified constellations, and due to its small size and faint stars, there are no known mythologies associated with Volans.

INTERESTING FACTS

Plancius based this new constellation on observations of the southern stars made earlier the same decade in Madagascar and the Indian Ocean by Pieter Dirkszoon Keyser and Frederick de Houtman (see **22. Chamaeleon**). On this trip, they apparently saw flying fish being chased by mahi-mahi (*Coryphaena hippurus*), which inspired this and the pursuing constellation, **33. Dorado**. These two were often shown running alongside the great boat constellation Argo Navis, before it was broken up into smaller pieces (see **17. Carina**).

88. VULPECULA, THE 'FOX'

Pronounced: 'vul-PECK-yoo-lah'
Short: Vul
Brightest star: α Vulpeculae or Anser (RA 19h 28m, Dec. +24°39')

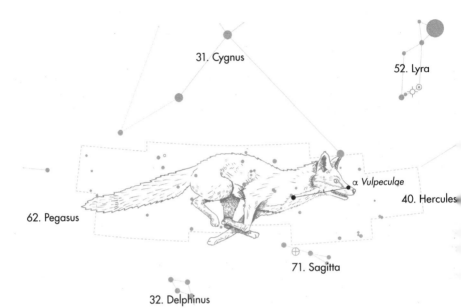

31. Cygnus

52. Lyra

α Vulpeculae

40. Hercules

62. Pegasus

71. Sagitta

32. Delphinus

In 1690, three years after the death of Johannes Hevelius, his wife and
fellow astronomer Elisabeth published a star catalogue based on the observations they'd carried out over the decades (see **45. Lacerta**). This included
Vulpecula cum Anser, 'a little fox with a goose'. The name was later abbreviated to just 'the Fox'.

WHEN AND WHERE TO OBSERVE

This is a relatively small and very faint constellation, and yet it's easy to find
because it sits between three much brighter stars that form what's known
as the Summer Triangle: Altair in **5. Aquila**, Deneb in **31. Cygnus** and Vega
in **52. Lyra**. Vulpecula is visible in dark skies in latitudes above −55°, most
prominently in September.

THE BRIGHTEST STARS

α Vulpeculae is the brightest star in the constellation, but it has an apparent
magnitude of only 4.4. It's known as Anser, the Latin for 'goose', which is
apparently held in the fox's jaws. It's an M-class red giant some 43 times
the diameter of the Sun.

OTHER BODIES

In 1967, British astronomer Jocelyn Bell discovered a relatively strong radio signal coming from somewhere in this constellation; the signal had a regular, quick pulse that repeated every 1.34 seconds. At first it was thought – not entirely seriously – that such a strong, regular signal must have been purposefully broadcast by intelligent aliens. The point from which the source emanated was called LGM1 after these 'little green men'.

Bell soon discovered another radio signal similar to this one, coming from another part of the sky. Fairly quickly, scientists worked out that these pulsing signals are the result of fast-spinning neutron stars radiating energy from their poles. If the neutron stars are angled the right way, the effect for us on Earth is like watching the spinning light of a lighthouse: it seems to wink on and off. Neutron stars that generate these regular pulses of energy are now called pulsars.

METEOR SHOWERS

There are no significant meteor showers associated with this constellation.

MYTHOLOGIES

As one of the more recently identified constellations, and due to its small size and faint stars, there are no known mythologies associated with this constellation.

INTERESTING FACTS

The Dumbbell Nebula (M27) is a large, bright planetary nebula – the first of its kind to be discovered, by French astronomer Charles Messier in 1764. In dark skies, it can be seen with good binoculars as a dimly glowing disc. The nebula is about 9,800 years old. Observation with a telescope reveals a double-lobed shape resembling an hourglass.

FURTHER READING

International Astronomical Union, 'The Constellations', https://www.iau.org/public/themes/constellations/.

Levy, D, Skywatching: *The Ultimate Guide to the Universe* (Collins, 1995).

Moore, P, *Philip's Guide to the Night Sky* (George Philips Ltd, 1991).

Ridpath, I, *Star Tales* (Lutterworth Press, 2018).

In-the-Sky.org provides guides to the night sky.

If you've enjoyed *The Art of Stargazing*, you might like *The Sky at Night: Book of the Moon* by Maggie Aderin-Pocock (BBC Books, 2018).

ACKNOWLEDGEMENTS

I would like to acknowledge all the people who have helped me with this book. To Simon Guerrier, for his dedicated editorial work and overall contribution to the book. I also want to thank Nell Warner, for her patient and dedicated support in editing, and Morgana Chess for her expert publicity skills. I also want to thank the team at Penguin Random House for all their support and guidance.

For Figure 1 – different interpretations of the stars in **60. Orion** – we drew on www.figuresinthesky.visualcinnamon.com/#sky-cultures. For indigenous Australian interpretations of **5. Aquila** we drew from www.assa.org.au/resources/aboriginal-astronomy/eagle-dreaming/, and for Alaskan interpretations of Sirius in **14. Canis Major** we drew from p. 23 of J.B. Holberg's *Sirius – Brightest Diamond in the Night Sky* (Springer, 2007). We consulted Dr Chris Naunton on the lack of archaeological evidence for Ethiopian queen **18. Cassiopeia**. For Tucano interpretations of **28. Corvus** we drew from escholarship.org/uc/item/7qn7t90d, and for ancient Greek interpretations of **31. Cygnus** we drew from penelope. uchicago. edu/Thayer/E/Gazetteer/Topics/astronomy/_Texts/secondary/ ALLSTA/Cygnus*. html. We consulted Dr Una McCormack, author of *The Autobiography of Mr. Spock* (Titan, 2021), on the location of Spock's home planet in **36. Eridanus**. Various dates have been suggested for the earliest known human remains found at Table Mountain in South Africa; for our entry on **53. Mensa** we drew from 'A mid-Holocene AMS 14C date for the presumed upper Pleistocene human skeleton from Peers Cave, South Africa' by Deano D. Stynder, Fiona Brock, Judith C. Sealy, Sarah Wurz, Alan G. Morris and Thomas P. Volman, a paper published in the *Journal of Human Evolution* in 2009. We also consulted A. Rousseau and S. Dimitrakoudis' paper 'A study of catasterisms in the "phaenomena" of Aratus', published in *Mediterranean Archaeology and Archaeometry* in 2006, and J.H. Rogers' paper 'Origins of the ancient constellations: I. The Mesopotamian traditions', published in the *Journal of the British Astronomical Association* in 1998.

INDEX

Page references in *italics* indicate images.

Andromeda 3, 22, 32–4, *32*, 35, 77, 78–9, 83, 86, 114, 142, *142*, *183*, 184, *185*, 186, *191*, 225
Andromeda Galaxy (M31) 23, *32*, 33, 34, 184, 226
Antarctic 19, 115, 179, 228
Antlia 35–7, *35*, 55, 74, *80*, 90, 91, 124, 134, *136*, 162, 165, 170, 171, 173, 189, 198, *198*, 201, 211, 222, *237*
Aphrodite 95, 97, 99, 192, 195
Apollo 97, 102, 127, 137, 176, 203
Apus 38–9, *38*, *45*, *88*, *90*, 115, *169*, 170, *173*, *181*, 220, *227*
Aquarius 24, 40–42, *40*, *43*, 44, 48, *71*, *85*, 86, *111*, 119, *119*, *183*, *191*, *194*, 195, 203, *210*
Aquila *40*, 43–4, *43*, *71*, 81, 97, 110, *111*, *131*, 159, *175*, *202*, 203, *204*, 212, *214*, 243
Ara *38*, 45–6, *45*, 81, 82, *96*, *171*, *181*, 207, 222
Aratus 53, 54, 81, 96, 110, 120, 122, 127, 132, 150, 175, 176, 184, 185, 192, 202, 205, 208, 210, 215, 233, 236, 240, 246; *Phenomena* 21, 75, 203
Arctic Circle 179, 203, 228
Arcturus 11, 26, 52, *52*, 53, 54, 159, 232, 239
Argo/Argonauts 74, 75, 93, 122, 128, 156
Argo Navis 21, 75, 76, 93, 196, 197, 198, 237–8, 242
Aries 24, 47–8, *47*, *85*, 152, *185*, *191*, *219*, 225, *225*
asterisms 3, 32, 41, 43, 58, 77, 83, 102, 105, 111, 115, 131, 183, 193, 204, 231
astrological zodiac 21, 23–4, 26, 40, 47, 48, 59, 70, 71, 126, 144, 151, 152, 175, 176, 177, 191, 204, 207, 219, 239
Auriga 49–51, *49*, *57*, 93, *126*, *155*, 156, *185*, *219*

Bayer, Johann: Bayer Designation 23, 45, 89, 148, 155–6, 182, 197, 208, 237; *Uranometria* 45, 129, 170, 228, 241
Betelgeuse *2*, 3, 11, 12, 26, *178*, 179, 180
binoculars 16, 27, 33, 39, 41, 46, 50, 60, 63, 95, 109, 127, 137, 151, 170, 197, 205, 244
black holes 13, 39, 56, 81, 84, 91, 104, 109, 141, 148, 188, 217; supermassive 56, 81, 91, 104, 141, 148, 217
Boötes 12, 26, 52–4, *52*, *62*, 63, 64, 69, *94*, *98*, *116*, *131*, 133, 159, *214*, *231*, 232, *239*
Brazil 69, 107, 230

Caelum 55–6, *55*, *92*, *114*, *121*, *134*, *149*, 189
Camelopardalis 34, 49, 57–8, *57*, 77, *83*, *116*, 155, 156–7, *165*, *235*
Cancer 9, 17, 24, 59–61, *59*, *68*, *126*, *136*, 137, *144*, *147*, *155*
Canes Venatici *52*, 53–4, 62–4, *62*, *94*, 143, 156, *231*
Canis Major 12, 15, 32, 39, 49, 54, 60, 63, 65–7, *65*, 68, 69, 75, 76, 81, 92, *92*, *149*, 150, 166, *166*, 180, 196, *196*, 213
Canis Minor 54, *59*, 68–70, *68*, *126*, *136*, 150, *166*, 180
Canopus 55, 74, 75, 76, 81, 92, 189
Capella 49, *49*, 50, 51
Capricornus 24, 40, *40*, *43*, 71–3, *71*, 80, 152, 163, *163*, 192, *194*, *204*
Carina 21, 55, 69, 74–6, *74*, *80*, 81, *88*, 92, 93, 128, *169*, 182, *189*, 196, 197, 198, 199, 237, *241*, 242
Cassiopeia 32, *32*, 33, *57*, 77–9, *77*, *83*, 84, 86, *142*, 143, *185*, 186, 225
Castor and Pollux 50, 60, 68, 120, 126–8, *126*, 180
celestial south pole (CSP) 39, 173, 174
Centaurus 14, 16, *35*, 46, 74, 80–82,

80, 90, 97, 105, 106, 107, 136, 151, 153, 154, 169–70, 169, 171, 184, 188, 205, 227, 237, 237

Cepheid variables 15, 22, 84, 115, 153–4

Cepheus 15, 22, 33–4, 57, 77, 77, 79, 83–4, 86, 93, 108, 115, 116, 142, 154, 186, 235

Cetus 33, 40, 41, 47, 79, 85–7, 85, 121, 124, 145, 184, 186, 191, 210, 219

Chamaeleon 38, 74, 88–9, 88, 107, 115, 129, 139, 140, 161, 169, 169, 173, 181, 187, 230, 241, 242

Chiron 46, 82, 120, 184, 188, 205

Circinus 38, 80, 90–91, 90, 114, 153, 169, 171, 172, 227

Columba 55, 65, 92–3, 92, 149, 189, 196

Coma Berenices 52, 62, 94–5, 94, 102, 144, 231, 239, 240

comets 8, 17, 20, 33, 50, 58, 72, 106, 117, 127, 132, 159, 179, 186, 220

Corona Australis 44, 45, 96–7, 96, 99, 204, 207, 222–3, 222

Corona Borealis 52, 96, 97, 98–100, 98, 131, 133, 214

Corvus 101–2, 101, 103, 104, 136, 137, 239

Crab Nebula (M1) 8, 219, 220, 221

Crater 101, 102, 103–4, 103, 136, 137, 144, 216, 239

Crux 39, 57, 70, 80, 81, 89, 92, 105–7, 105, 108, 115, 119, 129, 139, 140, 157, 167, 169, 174, 181, 187, 195, 228, 230, 241

Ctesias of Knidos 167, 168, 195

Cygnus 83, 108–10, 108, 116, 127, 142, 158, 158, 160, 183, 183, 243, 243

declination 30, 47, 115, 208, 217, 235–6

Delphinus 40, 41, 43, 111–13, 111, 119, 183, 202, 243

Dorado 8, 55, 114–15, 114, 134, 138, 161, 162, 189, 200, 230, 241, 242

Draco 52, 57, 83, 108, 110, 116–18, 116, 131, 131, 132, 133, 158, 160, 215, 231, 235

Dubhe (the 'bear') 22, 23, 232

Earendel (WHL0137–LS) 86, 145

equinoxes 30, 48, 106, 152

Equuleus 40, 111, 119–20, 119, 183, 190

Eratosthenes of Cyrene 95, 112, 127, 175, 184, 185, 203, 206, 208, 215, 226, 236

Eridanus 55, 85, 86, 121–3, 121, 124, 134, 138, 149, 178, 187, 188, 219, 229

Eudoxus of Knidos 44, 81, 110, 116, 120

exoplanets 16–17, 39, 41, 44, 46, 48, 81, 93, 110, 141, 168, 186, 195, 209, 232, 238, 240

Fomalhaut 86, 124, 183, 184, 192, 194, 194, 211

Fornax 85, 121, 124–5, 124, 187, 210

Galilei, Galileo 19, 60, 164, 180, 223, 232; Starry Messenger 180, 223

Ganymede 41, 44, 203

Gemini 24, 49, 50, 59, 60, 68, 68, 110, 120, 126–8, 126, 155, 155, 166, 166, 178, 180, 219

Gerard of Cremona 64, 154

Gilgamesh 76, 180, 220

Grus 129–30, 129, 140, 163, 181, 187, 187, 194, 195, 210, 229, 229

habitable or 'goldilocks' zones 36, 42, 60, 61, 82, 160, 190, 209

Halley, Edmond 76; Halley's Comet 179

Hercules (constellation) 43, 52, 82, 98, 98, 114, 116, 117, 127, 131–3, 131, 137, 146, 153, 158, 175, 185, 202, 202, 203, 206, 208, 214, 221, 243

Hercules (mythical character) 60–61, 127, 132, 133, 137

Herodotus 112, 188

Herschel, Caroline 78; Catalogue of Nebulae and Clusters of Stars 184

Herschel's Garnet Star 84

Herschel, John 162

Herschel, William 84, 137; *Catalogue of Nebulae and Clusters of Stars* 184
Hesiod 137, 188
Hevelius, Elisabeth 64, 141, 142, 143, 147, 155, 156, 157, 216, 243
Hevelius, Johannes 64, 87, 142, 143, 147, 155, 156, 157, 212, 216, 218, 243
Hipparchus of Nicaea 119, 120, 154
Homer: *Iliad* 65–6, 75, 180, 220, 233, 236; *Odyssey* 53, 180
Hondius, Jodocus 89, 129, 139, 140, 181, 187, 230, 241
Hooke, Robert: *Micrographia* 164–5
Horologium *55, 114, 121,* 134–5, *134, 138, 200,* 201
Houtman, Frederick de 89, 107, 115, 129, 130, 139, 140, 169, 181, 187, 188, 230, 234, 242
Hubble Deep Field 234
Hubble Space Telescope 14, 15, 78, 86, 167, 170, 195, 215, 232
Hubble, Edwin 22
Huygens, Christiaan 135, 223
Huygens, Constantijn 223
Hyades 9, 220
Hydra *35, 59,* 60–61, *68,* 70, *80,* 101, *101,* 102, *103,* 132, 136–7, *136,* 138, *144, 151, 153, 166, 196, 198,* 215, *216, 239, 239*
Hydrus *114,* 115, *121, 134,* 138–9, *138, 161, 173, 187, 200, 229*
Hyginus, Gaius Julius 54, 127, 192, 203, 226

Icarius 54, 69
Icarus 14, 86, *144,* 145
International Astronomical Union (IAU) 3, 4, 21, 22, 23, 24
Io 20, 221

Jupiter 19, 20, 60, 127, 141, 190, 199, 209, 238

Kaus Australis 204–5, *204*
Kepler, Johannes 115, 176
Kepler Space Telescope 110, 160
Keyser, Pieter Dirkszoon 89, 107, 115,
129, 130, 138, 139, 140, 169, 181, 187, 188, 230, 242

Lacaille, Nicolas-Louis de 35, 36–7, 55, 56, 74, 75, 90, 91, 124, 125, 134, 135, 137, 139, 161, 162, 165, 170, 171, 172, 173, 174, 189, 190, 196, 197, 198, 199, 201, 211, 222, 224, 237–8
Lacerta *32,* 64, *77, 83, 108,* 141, 142–3, *142,* 147, 155, 157, *183,* 212, 216, 243
Large Magellanic Cloud *114,* 115, 162, 230, 242
Leo 11, 14, 24, 58, *59,* 60, 86, 94, *94, 103, 136,* 144–6, *144,* 147, *147,* 148, 154, *155,* 207, 216, *216, 231,* 232, 239, *239*
Leo Minor *59,* 143, 144, *144,* 147–8, 147, *147,* 148, *155, 231*
Leonis 11, 144–6, *144*
Lepus *55,* 66, *92, 93, 121,* 149–50, *149, 166, 178,* 180
Libra 24, 63, *80, 136,* 151–2, *151, 153, 175, 207, 214,* 239, *239*
Lupus *80, 90, 136,* 151, 153–4, *153,* 170, *171, 207*
Lynx *49, 57, 59, 126,* 143, 144, *147,* 155–7, *155, 231*
Lyra 11, 15, *108,* 110, 112, *116, 131,* 133, 158–60, *158,* 243, *243*
Lyrae 11, 110, 158–9, *158*

Magellan, Ferdinand 39, 115
Magellanic clouds *114,* 115, 162, 230, 242
Mars 18, 19, 179, 208
Mensa *88, 114, 138,* 161–2, *161, 173, 241*
Merak 22, 23, 232
Mercury 18–19, 20, 123, 179, 189, 208
Messier, Charles 8, 220, 221, 244
Messier objects 8, 48, 72, 78, 84, 86, 90, 97, 102, 112, 117, 122, 125, 127, 130, 132, 135, 137, 139, 143, 150, 152, 154, 156, 159, 162, 163–4, 167, 170, 174, 182, 186, 188, 192, 195, 199, 201, 203, 208, 211, 217, 223, 232

Microscopium 163–5, *163*

Milky Way 33, 34, 43, 56, 90, 106, 109–10, 112, 115, 120, 130, 137, 141, 142, 148, 166, 167, 170, 182, 197, 199, 201, 204–5, 226, 230

Monoceros *65, 68, 126, 136, 149,* 166–8, *166, 178, 196*

Moon 6, 18, 19, 20, 25, 48, 85, 104, 143, 154

MUL.APIN 41, 145, 191, 205, 208, 220, 225, 240

Musca *38, 74, 80, 88,* 89, *90, 105,* 106, 169–70, *169*

nebulae 8–9, 26, 27, 44, 46, 69, 74, 81, 109, 122, 130, 184, 205, 217

Neptune 12, 20, 33, 73, 195

New Horizons space probe 23, 206

Norma *45, 90, 153,* 171–2, *171, 207,* 227

north celestial pole 4, 30, 57, 83, 116, 233

Octans *38,* 70, *88, 138, 140, 161,* 173–4, *173,* 218, *229*

Oort Cloud 20, 122

Ophiuchus 24, *43,* 131, *131, 151,* 175–7, *175, 204, 207,* 214, *214,* 215

Orion 2, *2,* 3, 9, 11, 12, *24,* 26, 60, 65, *65,* 66, 68, 85, *121,* 126, *126,* 143, 149, *149,* 150, 166, *166,* 178–80, *178,* 206, 208, 219, *219,* 223

Orion Nebula 9, 179, 180

Osiris 75, 150, 220

Pavo *38, 45,* 129, *140, 173,* 181–2, *181,* 187, *222,* 229

Pegasus *32,* 33, 34, 40, *40,* 79, 80, 86, *108, 111,* 119, *119,* 120, *142,* 183–4, *183,* 186, 191, *191,* 193, 194, *194,* 243

Perseus 9, *32,* 34, *47, 49, 57,* 77, 79, 86, 109, 184, 185–6, *185,* 199, *219,* 225

Phoenix *121, 124,* 129, *129, 138,* 181, 187–9, *187, 210,* 229, *229*

Pictor *55, 74, 92, 114,* 189–90, *189, 196, 241*

Pisces 24, *32, 40,* 41, *47,* 48, 73, *85,* 86, *183,* 191–3, *191,* 195

Piscis Austrinus *40,* 41, 86, 124, *129,* 130, *163,* 184, 192, 194–5, *194,* 210, 211, 241

Plancius, Petrus 39, 57, 89, 129, 139, 140, 156–7, 167, 168, 169, 181, 187, 195, 227, 228, 230, 241, 242

planisphere 27, 162

Pleiades 9, *178,* 180, 219, *219,* 220, 223

Pliny the Elder 120, 164

Plough or Big Dipper 3, 21, 22, 25, 26, 53, 63, 77, 145, 225, 226, 231, 235, 239

Pluto 20, 22, 101, 137, 206

Polaris (North Star) 5, 14, 25, 77, 117, *173,* 174, 232, 235–6, *235*

Polaris Australis (southern pole star) 173–4, *173*

Poseidon 33, 112, 120

precession 47, 48

Proxima Centauri 14, 16, 80, *80,* 81, 82

Ptolemy, Claudius 34, 51, 59, 60, 61, 62, 63, 66, 68, 72, 77, 81, 83, 84, 86, 95, 96, 98, 107, 110, 111, 116, 122, 126, 127, 132, 136, 143, 149, 152, 153, 154, 156, 158, 164, 183, 185, 194, 195, 196, 202, 205, 208, 215, 222, 226, 228, 236, 239

Ptolemy Cluster (M7) 86, 150, 186, 192, 203, *207,* 208

Puppis 21, *65,* 74, *74,* 92, *136, 166, 189,* 196–7, *196,* 198, 237, *237*

Pyxis *35,* 74, *74,* 91, *136,* 196, *196,* 198–9, *198,* 237

quasar 56, 104, 148

radial velocity method 16, 92–3

Regulus 11, *144,* 145

Reticulum *114, 134, 138,* 200–201, *200*

Rigel 26, *178,* 179, 180

right ascension (RA) 30, 184, 217

Sagitta *43, 111, 131,* 202–3, *202, 243*

Sagittarius 24, *43*, 71, 72, 96, 97, *140*, 163, 175, 204–6, *204*, 207, 212, *214*, *222*, 223

Sah 150, 180

Saturn 19, 20, 120

Scorpius 24, *45*, 46, 70, 81, 96, *96*, 151, *151*, 152, *153*, 154, 169–70, *171*, 172, 175, *175*, 176, 180, 204, *204*, 206, 207–9, *207*, 220

Sculptor *40*, 56, *85*, *124*, *129*, *187*, *194*, 210–11, *210*

Scutum *43*, 143, *204*, 212–13, *212*, *214*

'seeing' 25, 27, 28

Serpens 132, 136, 176, 214–15, *214*, 239

Serpens Caput *43*, *52*, 98, *98*, *131*, *151*, 175, *214*, 215, *239*

Serpens Cauda *43*, *175*, *204*, 212, *214*, 215

Seven Sisters 180, 220

Sextans *103*, *136*, 143, *144*, 174, 216–18, *216*

Sirius 12, 15, 60, *65*, 66, 75, 76, 81, 92, *178*, 180, 196, 213

Southern Birds 129, 181, 187, 188, 229

Southern Cross 80, 105–7, *105*, 108

Spica 26, 53, 102, 232, 239, *239*, 240

Spindle Galaxy (M102) 117, 217

stars: brightness, apparent 14–16; classes 10–13, *10*; composition 6–8; constellations as pattern and relationship imposed on 24, *24*; constellations of zodiac 23–4; defined 6; exoplanets 16–17, 39, 41, 44, 46, 48, 81, 93, 110, 141, 168, 186, 195, 209, 232, 238, 240; formation 8–13; labels 31; life cycle 9–13; movement, apparent 14; naming 21–3; observing 25–30; stellar radiation 13–14

states of matter, four 4, 7–8, *7*

Summer Triangle *43*, 110, 243

Sun: brightness 14; composition of 6–7, 17–20, *18*; constellations and *see individual constellation name*; exoplanets and 16; life cycle 10, 11, 12; stellar radiation and 13–14; system of 17–20, *18*

supernova 8, 9, 13, 66, 76, 78, 84, 120, 127, 141, 154, 156, 167, 176, 221, 238

Taurus 8, 9, 24, *47*, *49*, 66, 85, *85*, *121*, *126*, 144, 152, *178*, 180, *185*, *207*, 219–21, *219*, 223, 233

Teapot *204*, 212

telescopes 1, 5, 12, 14, 15, 16, 17, 19, 20, 21, 22, 23, 26, 27–8, 29

Telescopium *45*, 96, *96*, *140*, *163*, *181*, 182, *204*, 222–4, *222*

Tombaugh, Clyde 21–2, 137

Triangulum *32*, *47*, *185*, *191*, 225–6, *225*, 228

Triangulum Australe *38*, *45*, 70, *90*, *171*, 172, 227–8, *227*

Triangulum Galaxy/Pinwheel Galaxy (M33) 137, *225*, 226, 233

Tropic of Cancer 47, 72, 73

Tropic of Capricorn 72, 73

Tucana 115, *121*, 129, *129*, *138*, *140*, *173*, 181, 187, *187*, 229–30, *229*

Typhon 73, 192

Ursa Major 21, 22, 23, 25–6, *52*, 53, *53*, *57*, *62*, 63, 77, *94*, 100, *116*, *144*, 145, 147, *147*, 155, *155*, 156, 225, 228, 231–4, *231*, 235, 239

Ursa Minor 3, 235–6, *235*

Vega 11, 15, 110, 158–9, *158*, 243

Vela 21, 73, 237–8, *237*

Virgo 24, 26, *52*, 53, 54, 70, *94*, *101*, 102, *103*, *136*, *144*, *151*, *214*, 232, 239–40, *239*

Volans 74, *88*, *114*, 115, *161*, *189*, 241–2, *241*

Vulpecula *108*, 111, *131*, 143, *158*, *183*, 202, 240, 243–4, *243*

wain 3, 231, 233, 235, 236

White Tiger of the West 79, 128, 135

Willemsz, Vechter 89, 107

Wolf-Rayet star 66, 238

Zeus 44, 46, 50, 51, 54, 66, 72, 97, 110, 127, 132, 203, 221, 233, 236